Cover: Sunrise light on Minerva Terrace reveals the interplay of water flow, microbe growth and travertine mineral deposition. Colors are pigments from a thin veneer of living microbial mats. The meter-high brown stair-step travertine terracettes at the bottom were initially formed at lower flow rates. The top and middle 1-2 meter-high terracettes formed when increased flow changed microbial life and behavior. Pasta-like filamentous microbes, which grew out and over the top of underlying terracettes, were rapidly encrusted to form a rigid light beige travertine layer. Filaments draping down the edge are also quickly encrusted to develop tooth-like shapes similar to stalactites found in subterranean caves.

Opposite: Rays from the sun's reflection shimmer and highlight the scalloped edges of micro-terracettes in the morning steam at Grassy Spring.

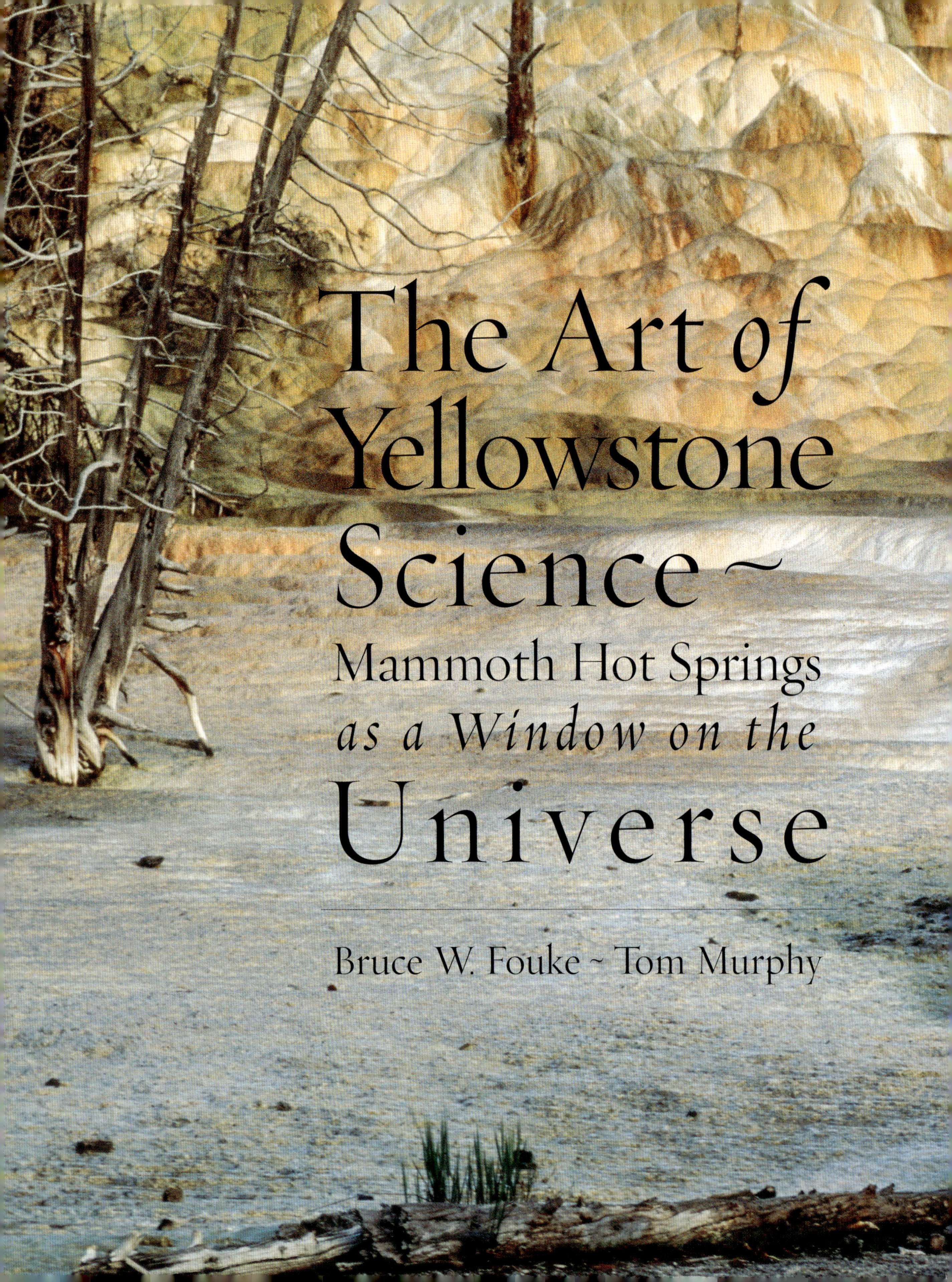

# The Art of Yellowstone Science ~

## Mammoth Hot Springs *as a Window on the* Universe

Bruce W. Fouke ~ Tom Murphy

Angel Terrace has flooded, killed and is entombing limber and lodgepole pine. Capillary action pulls mineral laden water up to a half a meter into tree trunks, creating white bases. The white streak running from the upper right down to the middle of the photograph is travertine deposited in an especially hot channel. The orange and brown mounded slope is also active, yet cooler and slower.

*Dedicated to the pursuit of truth and beauty in nature through science and art.*

Early morning sun backlights steam frosted trees and fog at Palette Spring. Microbes are being carried up into the atmosphere in the steam on this cold winter morning.

# Table of Contents


| | | |
|---|---|---|
| **Preface** | | **11** |
| **Chapter One** | **Wilderness Laboratory**<br>In a Golden Age of Scientific Discovery—<br>Why Mammoth Hot Springs?<br>Mammoth as Inspiration and Scientific Adventure<br>Confronting Complexity, Space and Time | **24** |
| **Chapter Two** | **Perspectives**<br>Science, Art and Everyday Life<br>Reductionism, Holism and a Middle Path<br>Inquiry in Science<br>Inquiry in Art | **47** |
| **Chapter Three** | **Dynamics**<br>Travertine Growth, Shape, and Composition<br>The Concept of Travertine Facies<br>Flowing Hot Spring Water<br>Geochemistry of the Hot Spring Water and Travertine:<br>A Perfect Storm<br>Diverse Ancient Microbial Life<br>Mammoth: The Unexpected Product of<br>Life-Earth Interactions | **77** |
| **Chapter Four** | **Emergence of Life**<br>A Family Tree<br>Evolution of the Tree of Life<br>Molecular Phylogeny<br>How Did Life Emerge?<br>Into a Microbial World | **142** |
| **Chapter Five** | **Deep Time**<br>Time Keepers<br>The Beginning of Time<br>Earth Time<br>Land of Fire and Ice | **187** |
| **Chapter Six** | **Human History**<br>Early Inhabitants<br>Euro-American Westward Expansion<br>Yellowstone National Park—Birth of an Idea<br>Photography in Nature | **214** |
| **Chapter Seven** | **Worldwide Laboratory**<br>Science, Art and the American West<br>Life in Extreme Environments<br>Mammoth as a Window on the Universe<br>Education In and About the Wild | **241** |
| **Wildlife Photography Gallery** | | **280** |

CREDITS

All photographs *(unless otherwise indicated)* ©Tom Murphy
All Tom Murphy photographs in this book available as archival giclée prints.
Crystal Creek Press, 517 West Geyser, Livingston Montana 59047

Photographs pp. 79, 83, 89, 92, 93 top, 96, 97, 98-99, 101, 123, 124, 126, 132, 133, 136, 137, 138, 139, 263, 266, 267, 275, 295 ©Bruce W. Fouke

Image pages: 38, 39, 40, 41 ©Tom Murphy and Kevin Caravati, Georgia Tech University; 70, 271 ©NASA Hubble Space Telescope; 76 ©Bing maps; 147 ©Jason Lindsay, College of Liberal Arts and Sciences, University of Illinois at Urbana-Champaign; 158 © University of Illinois Archives, University of Illinois at Urbana-Champaign; 191 ©NASA Wilkinson Microwave Anisotropy Probe, Timeline of the Universe; 192 ©NASA/Goddard/Arizona State; 215, 218-219 ©Collection of United States Library of Congress, Geography and Map Division; 220 ©Collection of Buffalo Bill Center of the West, Cody, Wyoming; 224-225, 227 (horseshoe) Courtesy of National Park Service, Yellowstone National Park; 259, 260 left ©Kurt Kwast, Molecular and Integrative Physiology, University of Illinois at Urbana-Champaign, shot SCUBA diving with Bruce W. Fouke; 261, 264, 265 ©Glenn Fried, Mayandi Sivaguru, and Bruce W. Fouke, Carl R. Woese Institute for Genomic Biology; 269 ©NASA Mars Exploration Rovers; 270 ©The Museum of Modern Art/Licensed by SCALA/Art Resource, NY; 274 ©Melissa Weymiller shot during Mammoth field course taught by Bruce W. Fouke

Objects on p. 14 and pp. 215-219, 222-223, 226-236 courtesy of Tom Murphy from his collection of Yellowstone National Park memorabilia.



First Edition
Published in 2016 by Crystal Creek Press
Livingston Montana 59047
www.crystalcreekpress.com



Book design by Adrienne Pollard
Printed and bound in China

ISBN 978-0-9973039-2-6

We wish to express our heartfelt appreciation for the unconditional and stalwart support of our families. Thank you to Bonnie for all of her thoughtful and careful editing up to the last minute as this book goes out the door. Adrienne Pollard skillfully and tirelessly completed the design and layout, assembling hundreds of photographic images and adeptly coordinating them with the scientific content. This is her 9th book project with Tom. Professor Emeritus Bruce Michelson from the University of Illinois Urbana-Champaign, a Mark Twain scholar by trade, imparted a spirited voice to the text through perceptive and meticulous editing. This was in part honed by his active participation in several of Bruce's Mammoth field courses. Professors Carl Woese and Abigail Salyers in Illinois Microbiology motivated, encouraged and guided the scientific process.

Partial support was provided by educational and public outreach initiatives through the NASA Astrobiology Institute and the National Science Foundation. The 100-year long stewardship and protection of Yellowstone by the National Park Service is the fundamental component that has made this project possible. We are deeply indebted to the staff of the National Park Service and their award of permit number YELL-03060, under which the research at Mammoth has been completed. Special thanks to National Park Service rangers Jeanne Johnson, Bob Fuhrmann, Ann Rodman, Christie Hendrix, Brian and Jo Suderman, Hank Heasler, Stacey Sigler, and Sarah Haas for their decades of colleagueship and support. Thermal imaging was completed in collaboration with Kevin Caravati at the Georgia Institute of Technology.

Many colleagues at the Carl R. Woese Institute for Genomic Biology made important contributions, including Glenn Fried and Mayandi Sivaguru, who invested a decade of work to complete advanced microscopy analyses and co-author many publications, and Nick Vasi and his communications staff who provided invaluable photographic and editorial support. Other contributors on the Illinois campus included Killivalavan Solai, who worked diligently to design and prepare the scientific diagrams; James Whitacre, an avid Tetons climber who completed the GIS cartography; and Jason Mock, Colleen Cook and their staff who filmed and produced our EMERGENCE OF LIFE MOOC.

Sincere thanks go to Bruce's advisors Merrill Foster, Brian Glenister, Bob Aller, Bill Meyers, Rich Reeder, Wolfgang Schlager and Walter Alvarez who blazed the trail of integrating cutting-edge science with dynamic education. Colleagues from around the world have played an integral role, with deep appreciation to Charlie Werth, Rob Sanford, Ed Feser, Isaac Cann, Nigel Goldenfeld, Gary Olsen, Vincent Bulone, Mark Vermeij, Martin Carrera, Gene Robinson, Elbert Branscomb and Julie Woolsey. Many remarkable postdocs and students have driven the basic research, with special thanks to George Bonheyo, Yiran Dong, Duncan Keenan-Jones, Mike Kandianis, Tom Schickel, Jinzi Deng, Raj Singh, Roy Johnson, Jim Klaus, Jorge Frias-Lopez, Alan Piggot, Amanda Oehlert, Ryanne Ardisana, Erin Murphy, Joe Weber, Jim Bruce and Annette Merkel.

We are grateful for the inspiration we have received from our thousands of students over the last 30 years. They saw things that we did not and awoke us to other possibilities. We are humbled and energized by the new perspectives and insights they have brought to us, and hope this galvanizes their own appreciation for the infinite possibilities of art and science.

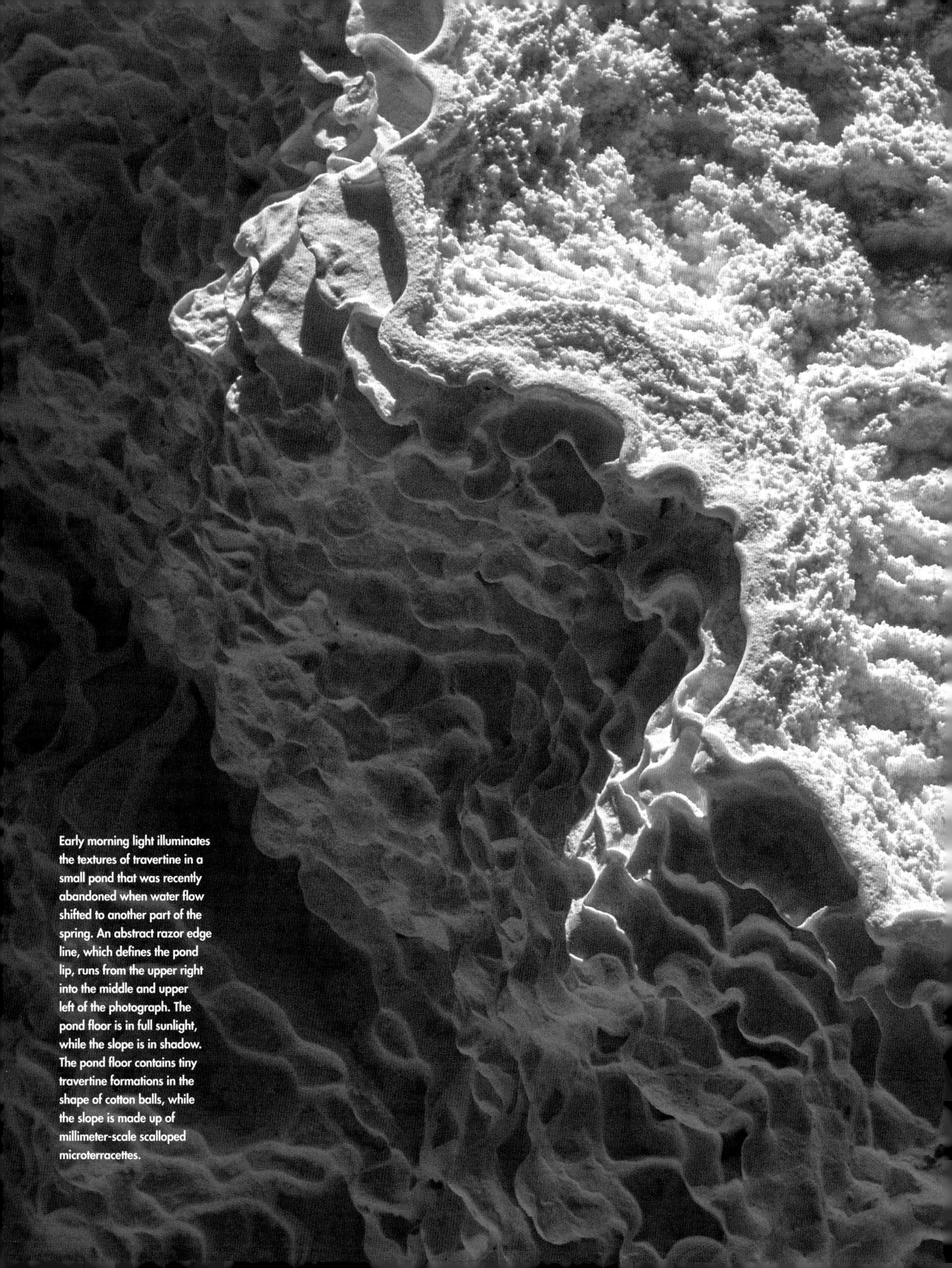

Early morning light illuminates the textures of travertine in a small pond that was recently abandoned when water flow shifted to another part of the spring. An abstract razor edge line, which defines the pond lip, runs from the upper right into the middle and upper left of the photograph. The pond floor is in full sunlight, while the slope is in shadow. The pond floor contains tiny travertine formations in the shape of cotton balls, while the slope is made up of millimeter-scale scalloped microterracettes.

There are two kinds of truth: the truth that lights the way and the truth that warms the heart. The first of these is science, and the second is art. Neither is independent of the other or more important than the other. Without art, science would be as useless as a pair of high forceps in the hands of a plumber. Without science, art would become a crude mess of folklore and emotional quackery. The truth of art keeps science from becoming inhuman, and the truth of science keeps art from becoming ridiculous.

— RAYMOND THORNTON CHANDLER
1888-1959

A complex multitude of sizes is shown here in the stair-step terracette formations on the slope at Palette Springs. The diverse green and orange colors are from thermophilic microbes, which inspired the name Palette.

ART AND SCIENCE BOTH ORIGINATE FROM THE SAME HUMAN DESIRE TO UNDERSTAND THE WORLD WITHIN AND AROUND US. Where did we come from? How did we get here? Why are we here? Where are we going? To seek answers, science requires inspiration from art, just as art yearns for understanding from science. This dynamic drives imagination and motivates curiosity.

This photograph was made in 1989, the year after the major Yellowstone fires. The bull elk is lying in a fresh green meadow contrasting with the black charred lodgepole trunks. This necessary cycle of fire and growth is what naturally happens in a healthy lodgepole ecosystem. Elk grow new antlers annually, and in the spring and in early summer they are covered by a soft fuzz called velvet. Notice the insect on his nose.

Truth, as expressed in nature, exists perfectly well without us. Photography can express truth more than any other media and capture complex unexpected interactions between life and Earth. Mammoth Hot Springs in Yellowstone National Park serves as a natural art and science laboratory to study the complexity, interrelatedness, relevance and power of all natural forces.

A photographic stereo-card made in 1878 by H.B. Calfee shows bathers at Canary Springs. Society has had drastically changing attitudes toward, and perceptions of, the value of Mammoth Hot Springs. Originally Mammoth was strictly recreational. Now we realize that thermal springs are extremely valuable for science and their recreational use is destructive in many ways. The higher purpose is to understand natural processes.

Canary is the largest active spring system in the Mammoth Hot Springs complex. Canary is a 20 m thick deposit of travertine that produces 1 ton per day of new travertine on its uppermost surface. The spring is reflected in a reservoir that was originally built as part of a water system for old Fort Yellowstone.

The rapid deposition of travertine at Palette Spring creates unstable rock formations that frequently collapse under their own weight.

Mammoth is a window on the universe, through which fundamental understandings of nature can be directly applied to other places and circumstances. Biological evolution is the framework for this combination of photographic art and science. Life originated billions of years ago in extremely hot aqueous environments similar to those found in Yellowstone. As a result, research to understand how ancient life originated on Earth and potentially throughout the cosmos has been focused on Yellowstone hot springs for decades.

Steam rising on a cold February morning from Canary Springs drifts into the branches of a fir tree in front of the sun. Tree branches cast shadows into the clouds of microbe-laden steam.

In the pages of this book, Mammoth Hot Springs photographic art is melded with the natural sciences to search for common laws of nature through the power of observation and a willingness to embrace the unexpected. A magnifying glass is first taken to observe the beauty and unexpected complexity of life in and around the thermal system at Mammoth. Time lines and historical events at Mammoth provide a context to appreciate the evolutionary Tree of Life established by Carl R. Woese. This new appreciation of nature at Mammoth is then applied to challenges faced by society, now and in the future.

A 5-month old grizzly bear cub is standing on its back legs and waving. Yellowstone lives through wilderness and wildness, and it provides wild creatures with the expanse of natural wild lands they need to survive and thrive. In so doing, the Park becomes an invaluable natural laboratory to study natural processes.

The truth of everything is in the truth of anything. Things happen and life exists for reasons that go far beyond our present understanding. Nature therefore stands as the most powerful and honest force in our lives, and human existence depends on living within the means of nature. Humanity's quest for a sustainable existence will define our present and future place in the universe. The art and science of Mammoth Hot Springs provides an unexpectedly insightful vantage point from which to envision this future. —BRUCE AND TOM

A winter sunrise illuminates the jagged glaciated features of Electric Peak.

Terracette pools containing green cyanobacteria at Canary Springs rapidly envelop a dead pine tree.

# WILDERNESS LABORATORY

## IN A GOLDEN AGE OF SCIENTIFIC DISCOVERY— WHY MAMMOTH HOT SPRINGS?

WHEN WE LOOK AT THEM CAREFULLY, patiently, the hot springs at Mammoth in Yellowstone National Park are transformed from a spectacle of color, water, and vapor into a window on the history and dynamics of the universe. Mammoth is a place to see powerful and important connections between deep time and human experience, among geology, microbiology, physics, and chemistry, as well as between the sciences and the arts. Art and science meld here because they are both essential and relevant. This extraordinary landscape and the forces beneath it move at their own cadence and will humble us as we try to observe them.

This shallow pool is located near the vent at Narrow Gauge. The pool is floored with hard white travertine perforated by gases produced by microbial metabolisms. Floating on the surface are filamentous *Sulfurihydrogenibium* bacteria undulating in the current.

To understand sources and relationships here, we must see not only in the ways that researchers approach an immense and wonderful natural scientific lab, but also in the artistic ways of Claude Monet, Vincent Van Gogh, Thomas Moran, and Ansel Adams. Watching the angles of light, the play of sun, shadow and color, we can come away with more than the pleasures of a momentary dazzle. We can, if we are lucky, achieve what William Wordsworth urged more than two hundred years ago:

> *... with an eye made quiet by the power*
> *Of harmony, and the deep power of joy,*
> *We see into the life of things.*

Though no set of photographs can transport us to Yellowstone, they can hold our gaze and center our attention on the vast and the miniscule, on the lasting and the ephemeral. Scientific study can ground these photographic observations, while strengthening and advancing our understanding of how Mammoth has come into being. Science deciphers what has happened here and what continues to unfold, while revealing the fabric and movement of the entire universe.

In nearly everything we do, we can feel the impact of our modern golden age of scientific and technological discovery. We have come to expect that the impossibilities of yesterday and today will become the easy routines of life tomorrow. For example, only a few years ago the chemical analysis *(molecular sequencing)* of entire strands of deoxyribonucleic acid *(the DNA genome)* required a decade of work by hundreds of specialists at a cost of millions of dollars. Today that kind of analysis can be completed in only a few hours by a handful of people at a tiny fraction of the previous price. With the ever-rising speed and plummeting cost of molecular sequencing, genome sciences have blossomed all over the world and now play a role in all sectors of society. With the advent of ultra-high resolution digital photography, the camera has taken on unprecedented importance in our quest to understand the life and meaning of Mammoth. As scientists we can now instantly work with rapid-fire images of objects thousands of times smaller than a living cell. With these enormous archives of information from new technology, we can collect sequencing and imaging data *(Big Data or Metadata)* that can be evaluated, "crunched," with computers of unprecedented speed and agility, some capable of a quadrillion calculations per second. This capacity for analyzing data allows science to attempt to understand complex processes in ways that were just recently impossible.

With regard to the life and natural history of Yellowstone, it is immensely important that we can now take many of our best research tools outdoors. Most of the microbial life on our planet cannot be grown *(cultured)* under controlled conditions in a laboratory. We, therefore, have to go out, find it and study it where it occurs. By applying new molecular sequencing and imaging technologies to all varieties of Earth environments,

A pond in the process of abandonment, with complex structures being revealed as they dry. The downstream face of the pond edges exhibit fluted microterracettes that range in size from 3–15 mm in diameter. Gas lift-off spired structures in the center of the photograph formed while previously submerged within the old pond.

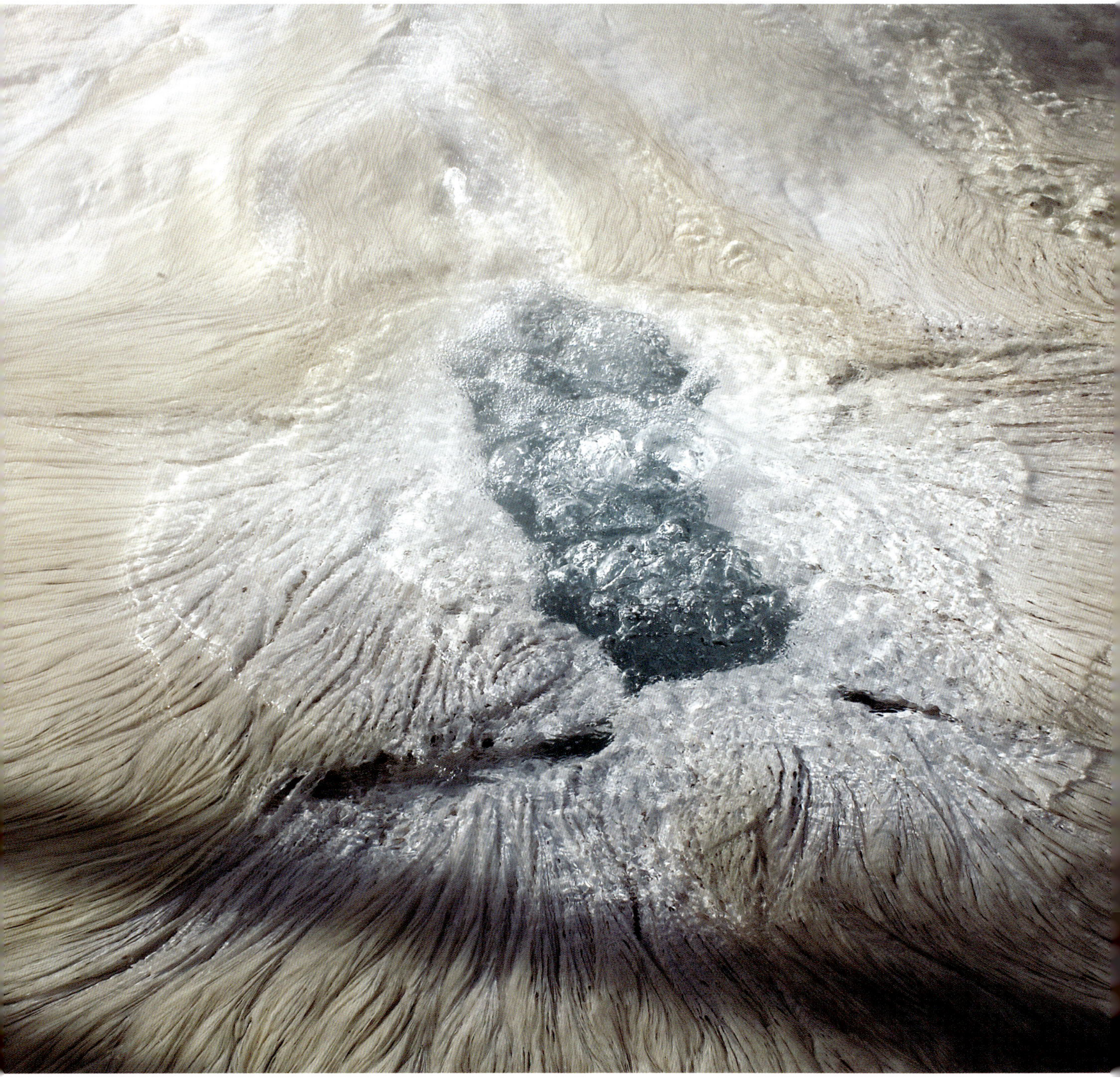

including the extreme conditions at Mammoth Hot Springs, we can detect and identify new forms of microbial life without laboratory culturing. This capability has opened up entirely new and unexpected insights into the composition, origin and distribution of life on our planet, and potentially everywhere else that life of any sort is possible. For instance, we can now estimate that there are 100,000 times more living cells inhabiting the Earth (about $10^{30}$), than there are stars and planets in the known universe. We can also understand that cellular life on our planet—the whole range of earthly life, including ourselves—began with microbes in hot-spring environments that in many regards were much like the ones that now thrive at Mammoth. These microbes evolved and diversified over immense stretches of time, in response to changing physical and chemical conditions, even when these changes were sudden and catastrophic. Today, it is estimated that more than 60% of all the life on our water-rich planet is microbial, with most of it dwelling within rocks buried deep in the Earth's crust. Hidden beneath our feet, we have a lot of company.

Microbial life is the oldest, largest reservoir of genetic diversity on our planet, dwarfing the biodiversity of all the plants and animals we can see with the naked eye that dwell in rain forests, coral reefs or the world's oceans. Microorganisms called cyanobacteria, which are not plants at all, but their distant primordial predecessors, invented photosynthesis and oxygenated the oceans and atmosphere of the Earth starting around 2.5 billion years ago (Bya). Still dominating global photosynthetic activity on our planet, microbial life helps drive many other processes, biological and geochemical, that are crucial to the healthy functioning of the global environment. In fact, the human body, which is at least 95% water, is made up of about 90% mammalian cells (by volume). Yet, because each microbial cell is nearly two orders of magnitude smaller than each mammalian cell, microbes (the ***microbiome***) make up 90% (by count) of the total number of cells comprising the human body. If microbes disappeared tomorrow, humans would cease to exist, but if ***humans*** disappeared tomorrow, microbial life on our planet would continue to thrive.

Opposite: $CO_2$ is degassing from the water rolling out of the Narrow Gauge vent. A dynamic hydrologic transition takes place from the vent into the apron and channel, creating a niche for filamentous *Sulfurihydrogenibium* bacteria, one of the most ancient and rapidly growing microorganisms on Earth. They metabolize dissolved sulfide and produce yellow-green elemental sulfur.

Unexpected new insights like these, on the elegant interconnectedness of life and the physical history of Earth, have required the sciences to integrate and work together as never before. For example, Geobiology and Astrobiology have emerged as important new disciplines to study complex physical, chemical, biological and even social interactions in nature, both on Earth and throughout the universe and through vast amounts of space and time. Along with cross-disciplinary thinking, this kind of work requires a combination of field and laboratory experimentation, as well as mathematical synthesis. The goal is to understand the unifying physical, chemical and biological themes of a natural system like Mammoth Hot Springs. How does it work on every level, from the single microbial cell to the entire ecosystem, and how do these components and dimensions interact? We also want to understand how these amazing environments evolved by learning to accurately read the fossil record of life laid down in the distant geological past. We are just beginning to learn how Mammoth came into existence, how its vitality and scope can best be conserved, and even how its secrets contribute to the well being of humankind.

## MAMMOTH AS INSPIRATION AND SCIENTIFIC ADVENTURE

Yellowstone National Park contains the world's largest assortment of hot springs, geysers, mud pots, and gas and steam vents (*fumeroles* and *solfataras*). All of these features emanate from the Yellowstone supervolcano, an underground sea of molten rock (*magma*) sixteen kilometers in depth with a total volume of 56,000 km$^3$ that could fill the Grand Canyon more than 14 times. The last cataclysmic eruption of this supervolcano, 640,000 years ago, spewed ash, rock, poisonous gas and destruction across North America from the Midwest to the Pacific Coast. The resulting caldera is a crater so wide and diversely featured that it is best and most fully comprehended from satellite imagery. On the ground, the supervolcano makes itself known to Yellowstone visitors by the presence of more than 12,500 hydrothermal features and regular earthquake activity.

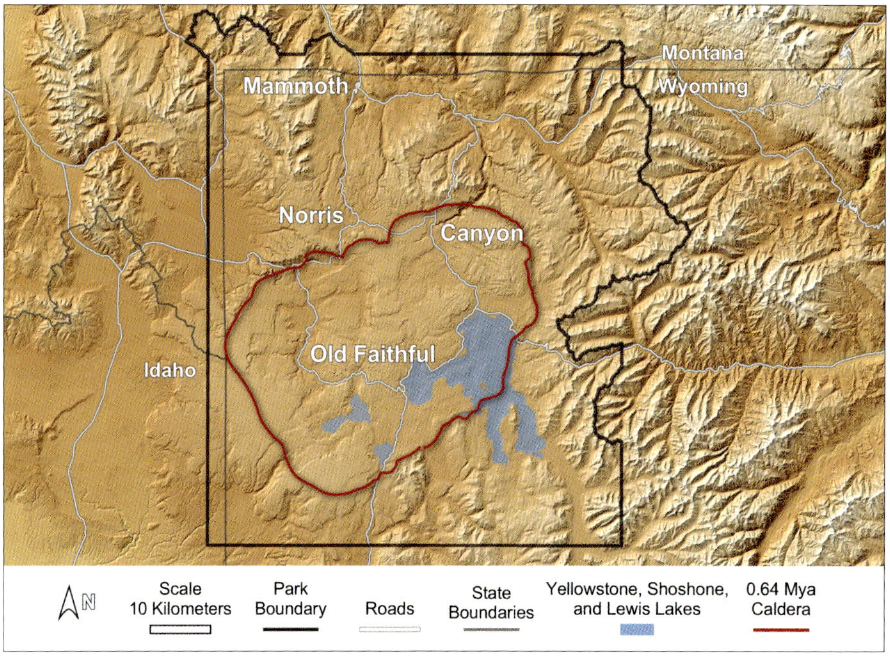


Topographic shaded-relief map of Yellowstone National Park showing the outline of the 640,000 year old supervolcano caldera *[from Lowenstern and Hurwitz, 2008]*.

Because Mammoth Hot Springs lies beyond the northern edge of the caldera, the water that reaches the Park here as rainfall and snowmelt undergoes a long and complex hydrologic adventure that is much different from water closer to and within the supervolcano caldera. The journey, which takes thousands of years to complete, progresses from the Gallatin Range, to deep within the Earth and ultimately back to the surface at Mammoth. In that long expanse of time the chemical changes the water undergoes are crucial to determining the geology and hydrology of the region. Moving down into the subsurface through the Swan Lake and Reese Creek faults, the water reacts at temperatures of over 100°C with

Travertine texture, in the lower half of the photograph, in the form of microterracettes precipitated under hot-spring water moving left-to-right.

Mississippian-age Madison Group limestone rock deposits. The water returns to the surface charged with Ca, Mg and $CO_2$ gas, thus permitting a rock called ***travertine***, a kind of $CaCO_3$ limestone, to be directly deposited from the flowing spring water at the surface of Mammoth. Constantly accumulating, layer upon layer, these deposits create and transform the landscape of the entire area. Because the deeply buried Madison limestone was originally laid down on the seafloor of oceans about 350 million years ago (Mya), the water emerging from vents at Mammoth has a chemical composition similar to seawater—albeit at a temperature of 73°C with neutral acidity ($pH = -\log[H^+] = 6$). The water emerging from Mammoth has thus gone through a long epochal journey of thousands of years, originally falling from the skies, seeping deep into the rocks of the crust, and rising back to the surface again. Yet once the water flows out from the spring vents, each only a centimeter or two in diameter, travertine is deposited at what seems to be impossibly fast rates that reach 5 millimeters in a single day. This is millions to billions of times faster than limestone deposition in other common Earth environments such as oceans, caves and lakes.

For the scientist working in the field at Mammoth, this is an enormous stroke of luck. To study processes by which the surface of the Earth transforms so quickly in mass and quality, means that experiments can be completed at Mammoth in just a few days. Compare this to the impossibly long durations that would be required in other natural environments. Mammoth is one of the two largest sites of active, travertine-depositing hot springs on the modern-day Earth, second only to those at Pamukkale, Turkey.

Right: Map of Mammoth Hot Springs, Wyoming, and Gardiner, Montana, at the northern margin of Yellowstone National Park.

Opposite: Dramatic lighting at Palette Springs highlights the rapid outward growth of filamentous bacteria over a pond as the volume of water increases. The resulting rock structure of streamer Apron and Channel travertine draping over Pond travertine is one of the most common and beautiful features seen from the boardwalk.

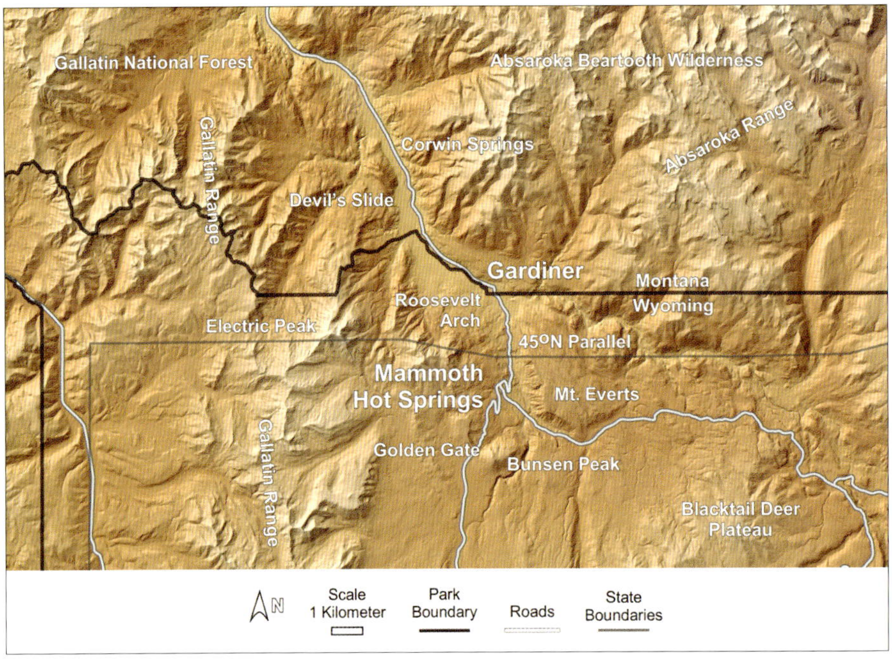


Travertine depositing hot-springs have been active in Mammoth for thousands of years, while just up the road in Gardiner, Montana (at the north edge of the Park) springs were active for tens of thousands of years. There they have created a layered sequence of deposits from the Holocene and Pleistocene epochs, hundreds of meters thick. These were interrupted for several millennia when the kilometers thick Pinedale glacier covered the whole region in ice. The water chemistries, temperatures and microbial communities responsible for depositing the ancient travertine at Mammoth (extending back to 8,000 years before present) and Gardiner (back to 39,000 years before present) have been reconstructed by direct comparison with the active flowing springs and their modern travertine on the top of Mammoth.

Preservation efforts by the U.S. National Park Service have maintained Mammoth in its natural healthy state since 1872. This is a rarity among travertine depositing hot-springs around the world, as most of them have been either heavily impacted by human use for centuries, or they are remote and difficult to reach. The geographic position of Mammoth, close to the northern boundary of Yellowstone, makes it readily accessible via roads, trails and boardwalks that are maintained year round. As a result, Mammoth is one of the most visited thermal features in the Park. But among these millions of annual visitors, very few understand that Mammoth is a place of "constant change," where new springs continually erupt, or stop their flow and shift to some other spot within Mammoth, over very short periods of time. The quiescent vents are scattered around and can give an impression that the Mammoth thermal system is shutting down, when it definitely is not.

This coral-like structure nestles in an abandoned pond. Indian rice grass has been encrusted with travertine that is chemically and mineralogically similar to the branched skeletons of marine corals.

## CONFRONTING COMPLEXITY, SCALE AND TIME

A visit to Mammoth Hot Springs is like stepping onto another planet. The treeless, sulfur-laden landscape of Palette, Canary, and the other famous sites here offers a myriad of multi-scaled travertine terraces, with thin veneers of flowing water and colorful microbial communities thriving just below the surface. To make real progress in understanding what we see at Mammoth, we need a method to approach this kind of extreme visual complexity. These are indeed heady encounters with phenomena that challenge our everyday assumptions about scale, change, and the passage of time. Sorting all this out has required extensive fieldwork, with many visits and observations done every year for decades, so that water flow rates, travertine growth, and microbial activity can be accurately measured and tracked. These measurements done over time *(time-series analyses)* have told us much about how these hot springs came into existence, and how to predict what will happen here in the future.

Before any scientific research can be done, the first full day of each fieldtrip is dedicated simply to looking at, sketching, photographing and contemplating the current state of each spring to be studied. Over the years, these initial observations have proven essential to achieving new insights and detailed hypotheses about the workings of the Mammoth thermal system. Recently, additional ways to "see" Mammoth and the Yellowstone region have been developed, providing perspectives that scientists only a few decades ago would not have thought possible. For instance, high definition satellite imaging (digital elevation maps and other remote sensing tools) now permits large-scale tracking, evaluation and contextualization of the entire Mammoth system. An invaluable addition to this is thermal infrared photography as a means of observing nature at Mammoth, allowing the observation and measurement of energy spectra normally invisible to our eyes. In the early 20th century it was confirmed that light acts as both a particle and a wave. Thermal infrared light is a form of electro-magnetic radiation generated by heat, which has a wavelength longer than that of visible light. Thermal infrared photography opens unseen and unexpected worlds in the hot-to-cold extremes of Mammoth and Yellowstone. These technologies permit time-series analyses at a wide range of small to large scales, permitting the physical, chemical and biological changes in many springs to be tracked simultaneously.

We have also learned to make use of a "Powers of Ten" strategy for gathering and collating data at Mammoth. This approach will be referred to repeatedly in upcoming chapters, but here is an example of how this strategy helps us to sort out a very complex natural setting. Bison like to shelter each winter from the bitter cold around the hot springs of Mammoth. As they graze and wander in and around the springs, many invisible interactions are taking place, such as hot-spring microbes going for a ride on the bodies of the animals, as well as in the steam. The vast supervolcano, encompassing everything we can see here, exerts its thermal influence constantly in any number of ways. How can we sort this out and keep a sense of proportion as we seek to understand these inter-relationships, when the scales we have to consider range from the extremely large to the extremely small?

Rainbow over Palette Springs and Devil's Thumb.

For instance, our measurements would necessarily include: (i) the average diameter of a hot-spring bacterium at Mammoth (1 micrometer or 1/1,000,000$^{th}$ of a meter); (ii) the average height at the shoulder of an adult bison (180 cm or 1,800,000 times the size of the average bacterium); and (iii) the diameter of the Yellowstone supervolcano caldera (72 km or 72,000,000,000 times the size of the average bacterium). Equally impressive are the time scales that encompass these interactions, including: (i) the time needed for one molecule of $CO_2$ to be consumed by a hot-spring bacterium at Mammoth (less than 1 nanosecond or 1/1,000,000,000 of a second); (ii) the age when developing bison are mature enough to produce a calf (about 3 years or 94,608,000 seconds); and (iii) the length of time since the Yellowstone caldera formed (640,000 years or 20,183,040,000,000 seconds).

Top: Overflow from Rainbow Pool in Black Sand Basin flowing into Iron Spring Creek. The bright orange and red show the flow of the hottest water, while the greens and blues reveal heat dispersion in the coolest water. The blue silhouettes are of dead pine trees.

Bottom: An explosion of heat in the center of Palette Spring, seen by the thermal infrared camera as white and yellow, dissipates into the sky as purple steam. Fingers of hot water appear red and purple as they drain down and cool into the black of the cold underlying rock.

The similarity in temperature at which complex life thrives in the core of a bison and Canary Springs is simultaneously revealed by this thermal infrared image.

Spring water cascades down the face of terracettes in the Proximal-Slope, cooling as it descends. The dark blue "dancing bear" in the center is actually a cold piece of travertine.

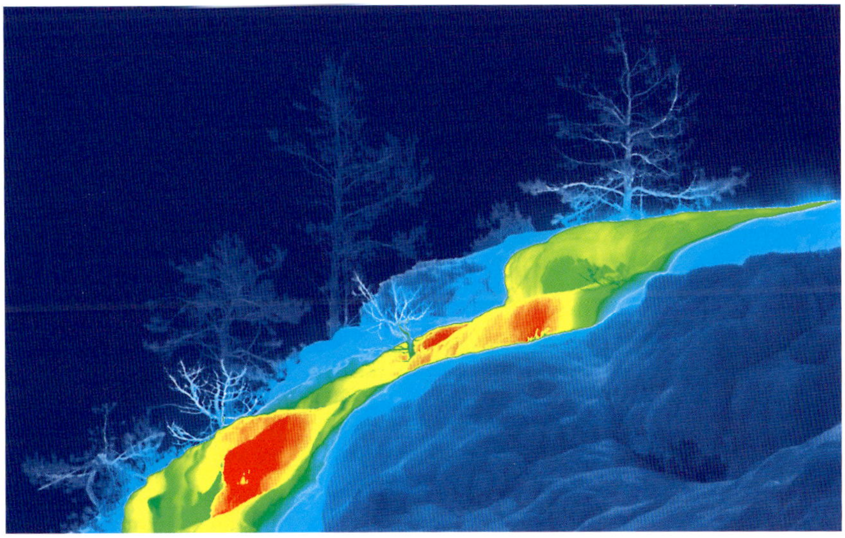

Top: An active channel at Palette Springs flows through the cold quiet blue of abandoned travertine, limber pine and sky.

Bottom: A thermal infrared photograph depicts hot water as black and vertical dark blue curtains that cascade out over the Palette Spring stepped terracettes. Background heat exists in the underlying travertine and surrounding trees.

As a way of thinking clearly about all this, we turn to exponential notation of the number 10. In this framework, the number $1 = 10^0$, $10 = 10^1$, $100 = 10^2$, $1000 = 10^3$, $1{,}000{,}000 = 10^6$, and so on. The Powers of 10 system uses exponents to make comparisons over enormous scales of space and time. As a result, something 1000-times ($10^3$) larger is described as being "three Powers of 10 larger", with each 10-fold difference referred to as "one order of magnitude." In the example of bison wintering at Mammoth Hot Springs, we have to think about a dynamic range of eight Powers of 10 in spatial dimensions, and eleven Powers of 10 in temporal dimensions. The Powers of Ten strategy does more than simply clarify comparisons. It provides a perspective for understanding deeper relationships, recurrences, resemblances, and counter-intuitive relationships that the vast differences in length scale and geological eras might otherwise cause us to miss. When we think about physical length, the 1 micrometer average diameter of a microbial cell (one micron, or one thousandth of a millimeter) can therefore serve as a unifying scale, across which the travertine, the flowing water, and the microbe components of the Mammoth system all interact.

Right: A bison bull uses fossil travertine at Soda Butte in the Lamar Valley as a windbreak on a cold, sunny January day. His steamy exhalations are visible in front of the shadow of his body. Soda Butte, Terrace Mountain and the Hoodoos are the only other major sites of travertine deposition in Yellowstone outside of Mammoth Hot Springs.

Opposite: Powers of Ten space and time composition of Mammoth Hot Springs—from microbes to bison to glaciers and the super-volcano.

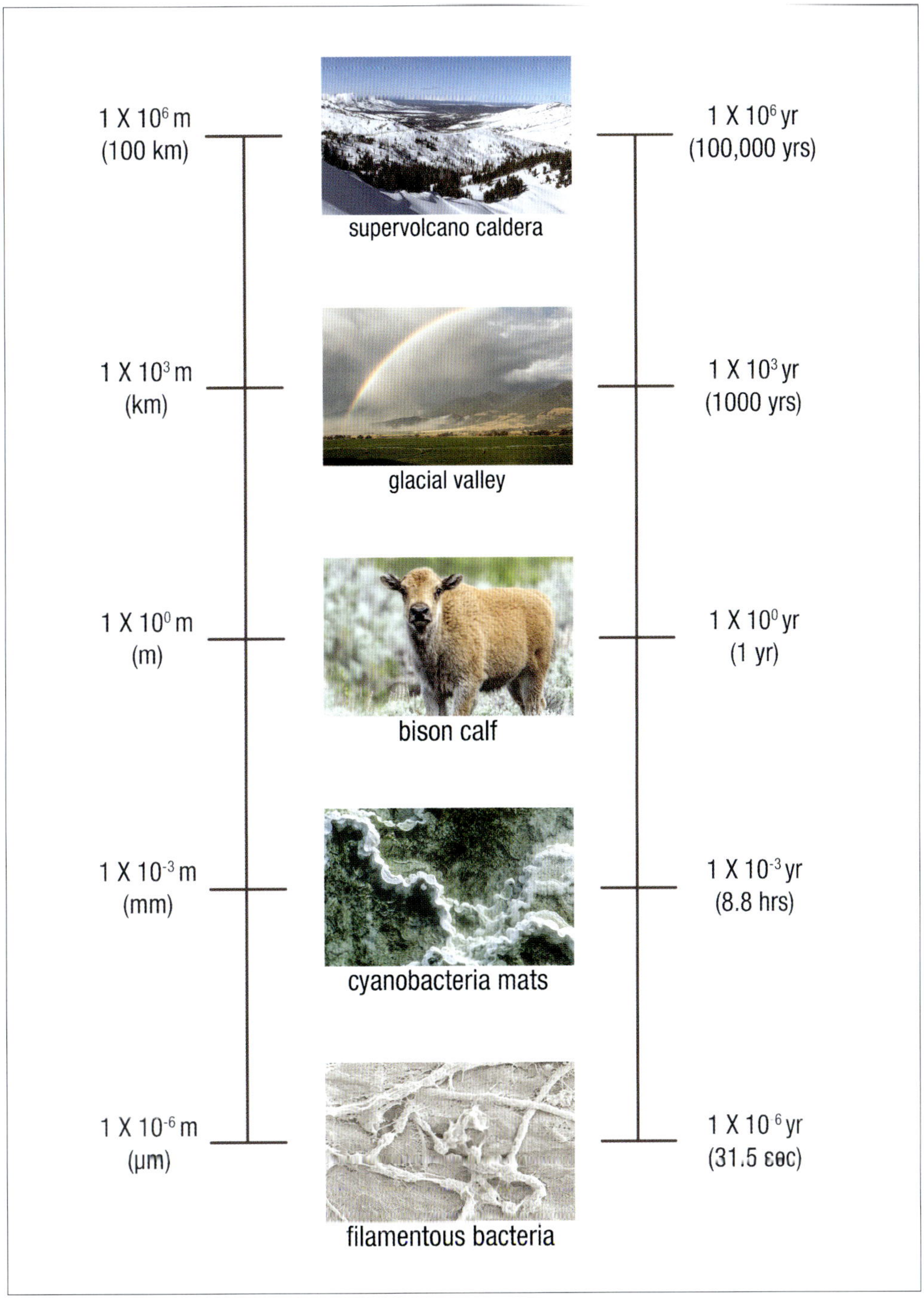
1 X $10^{6}$ m
(100 km)
1 X $10^{6}$ yr
(100,000 yrs)
supervolcano caldera
1 X $10^{3}$ m
(km)
1 X $10^{3}$ yr
(1000 yrs)
glacial valley
1 X $10^{0}$ m
(m)
1 X $10^{0}$ yr
(1 yr)
bison calf
1 X $10^{-3}$ m
(mm)
1 X $10^{-3}$ yr
(8.8 hrs)
cyanobacteria mats
1 X $10^{-6}$ m
(μm)
1 X $10^{-6}$ yr
(31.5 sec)
filamentous bacteria

Electric Peak stands near the northwest corner of Yellowstone. Tom skied to the top of Electric Peak one March day. While standing on top of the peak, he put his arm into the air and touched 11,000 feet above sea level (3,353 m). From there, he could see the full length of the Yellowstone ecosystem—from the Tetons in Wyoming to the Crazy Mountains in Montana. This is an expanse that is one order of magnitude larger than the supervolcano caldera. This rugged peak sculpted by glaciers is home to grizzly bears, wolves, mountain lions and wolverines, key species that occupy true wilderness.

A young bull elk uses fresh travertine to remove antler velvet. This results in the transportation of a multitude of bacteria to other sites in the Park.

# PERSPECTIVES

## SCIENCE, ART AND EVERYDAY LIFE

YELLOWSTONE NATIONAL PARK is a place of thrilling extremes. Atop and surrounding the supervolcano caldera, the plateau is a panorama of wilderness, wildlife, and hydrothermal features teeming with microbial diversity and action. Mammoth Hot Springs, uniquely beautiful and complex, demands a blend of art and science for us to grasp its truth and understand the rich complexity behind what a single glance can take in. As a process of rational thought, science is propelled by the kind of open, creative engagement that we commonly associate with the arts. What we choose to study in the natural world is often inspired by a sense of wonder and an awareness of the endless possibilities that lured great painters into the wilderness even before the advent of modern-style scientific study. All human senses and intellectual faculties arise from a need to understand nature. It thus follows that art and science together help us satisfy the imperative to know the world in which we live.

The visual arts, including photography, frame the natural context, helping us to concentrate our attention. Science incorporates the array of conceptual tools and procedures through which our observations and speculations about nature are verified, via reproducibility, logical interpretation, synthesis and making accurate predictions about the future.

The ultimate test of any scientific work is whether the conclusions can make accurate predictions. When we integrate art and science, we can reconcile the systematic and the personal in the act of observation. Though there can be no standard of "reproducibility" in painting or the art of photography, as there is in the realm of scientific experiment, we can scrutinize and appreciate paintings and photographs for how well they capture and tell the stories that science seeks to unfold. Equally valuable, art can also turn our scrutiny in an analytical direction we had not thought of before. Here are some cornerstone perspectives that help us to meld the artistic and scientific approaches.

Perhaps everyone in the modern world is at heart a functioning "Everyday Scientist," just as most of us are also "Everyday Artists." As we move through the world, we appreciate the beauty around us, and we speculate, at least on the fly, about what basic principles of light, chemistry, and biology bring it about. That kind of ordinary, universal inquiry goes on everywhere in the world, as humans deploy their powers of observation to calibrate their life experience as a means to survive, function and find

Right: Standing in a tree near its nest this bald eagle was warning a crow to leave the area. The crow must have had a nest in the neighborhood too because it harassed the pair of eagles when they went back and forth with food for their eaglets. Either species could predate on the other's nest, so the protective dance for both sets of babies went on all season.

Opposite: At Steamboat Point, the supervolcano caldera bisects the east shore of Yellowstone Lake. The Red Mountains, seen in the distance, stand at the south edge of the largest volcanic crater known on Earth, which formed some 640,000 years ago.

pleasure in our existence on Earth. Consider, for example, all the observations and surmises that might go into a simple decision out in the woods—where and how to cross a frozen stream to avoid falling into the frigid water. With reason and intuition, working logically and also aesthetically, we take in everything we see and hear and feel. This includes the look of the snow and ice, the possible effect of the sunlight and the heat of the day, the sound of the stream beneath the ice, water bubbling up where the surface is broken and the possible footing afforded by snow and ice-covered stones. To successfully accomplish our task, we incorporate how we feel in this rich environment, how strong and agile we are at that moment, our physical wherewithal for adapting to big steps, quick shifts, and rude surprises.

In the natural world, what catches our attention is often something new to us, strikingly different from what we have encountered before. For example, the unique patterns of terraced travertine rock deposits at Mammoth, with their multitude of sizes, shapes and colors, make them irresistible to look at and probably unlike anything we have seen anywhere else. The everyday scientist, struck by what he or she sees here, may return to the same vantage point again and again, to try to know more about how this complex natural system actually works. In the process, comparisons will suggest themselves and unexpectedly emerge. In the sciences, this common pattern in our thinking is summed up in the phrase "analogy identifies anomaly." That principle was first recognized long ago by working researchers and poets who noticed and celebrated its truth as well. Here are some lines from the American poet Richard Wilbur, probing the natural world by means of metaphors and similarities:

Opposite: Recent increases in outflow velocity at Grassy Spring created channels of high temperature water that cut down into lower temperature slope environments. Scale is deceptive. This looks like an aerial photograph of a river delta, when it is actually just 2 meters across.

Left: Low angle sunlight from the right partially shades shallow microterracettes that are forming on the Proximal-Slope Facies. Drying dark brown to black microbial mats are left behind after the water has shifted course and drained from these small ponded environments.

Right: A juvenile killdeer is stretching its wings while standing in thermal water. Steaming and splashing water precipitates traces of white aragonite on the feathers and down of this little bird. Unwittingly these birds carry travertine rock around on their bodies.

Opposite: This lodgepole pine was killed and is being entombed by encroaching Proximal-Slope Facies deposits at Palette Spring. Depending on the activity of the spring, this tree could be completely buried in a matter of years.

*Closer to making than the deftest fraud*
*Is seeing how the catbird's tail was made*
*To counterpoise, on the mock-orange spray,*
*Its light, up-tilted spine; or, lighter still,*
*How the shucked tunic of an onion, brushed*
*To one side on a backlit chopping-board*
*And rocked by trifling currents, prints and prints*
*Its bright, ribbed shadow like a flapping sail.*
*Odd that a thing is most itself when likened...*

This is how we look at new things, to understand them and to probe their depths. The greater our accumulation of life experience and learning, the stronger the basis from which we can make comparisons. For pioneering analysis of the new, as well as for deeper understanding of the known, progressive science focuses on anomalies. Successful and meaningful science must also be able to make consistently accurate and reproducible predictions. In the case of crossing that frozen stream, the everyday scientist draws from a wide array of previous experiences. And if he or she falls in, well, that provides the anomaly, the newly added experience from which to better negotiate the natural world in the future.

Bison cow and young calf walking across the base of the Proximal-Slope Facies travertine at Canary Springs.

## REDUCTIONISM, HOLISM AND A MIDDLE PATH

In studying complexity, one classic approach is called ***reductionism,*** which is an opposite strategy to ***holism.*** The Industrial Age has ushered in a time when mechanical principles and analogies were applied not just to manufacture but also to just about everything else. Reductionism works on an assumption that when powers of detailed observation join forces with available technology, a complex entity can be understood from knowing all of its individual components. For instance, if you were a strict reductionist scientist and a fuel injector on your car needed repair, you might say to yourself, "All right: to solve this problem, I will remove the injector unit, take it apart and spread all of its pieces on my workbench, inspect them all individually, swap out the defective ones, and put it all back together." Though such an approach might ultimately get you back on the road, you might be leery of all the time and effort required to break the system down like this and meticulously rebuild it. There is also a possibility that in the process some exotic adjustment or calibration might be lost, requiring you to start the business over again. Conversely, an ardently "holistic" researcher might say, "I'm going to examine the entire car now in its current state, all as one. I'll start with the front bumper and work all the way back to the tailpipe. I'll look carefully at the injector, of course, but I won't necessarily take it apart—not just yet. First, I'll want to know how the whole vehicle interconnects and normally operates, as a step to getting it going again."

Opposite: A white by-pass channel snakes its way down through the green, brown and orange pigmented microbial mat pigments of the Proximal-Slope Facies at Canary Springs.

Left: Meter-long light beige tufts of *Sulfurihydrogenibium yellowstonense* bacteria grow from multiple closely spaced vents on Main Terrace. The water flows in shallow turbulent sheets to lay down several mms per day of fettuccine streamer travertine textures.

Obviously, an approach like that also raises worries, that too much time will be spent on matters that will not help us see and solve the problem at hand. For all sorts of common-sense reasons, therefore, a mixed approach, a "middle path," is often the best recourse. We decide to look at one issue, with an awareness that other units or systems are also in the mix and may play an important role in how things function. In our example, the fuel lines to that injector, perhaps, or the ignition system, or the timing, the pollution-control connections, or there might be other things to consider to help make the car run smoothly. The middle-path approach bears in mind that "the whole is greater than the sum of its parts," while seeking to know how the most *relevant* parts work and how they contribute to the entire system.

Right: A dynamic interplay between life and hot-spring flow constantly unfolds during development of the Pond and Proximal-Slope Facies at Canary Springs. White by-pass channels shuttle hot water down into the cooler terracettes, as the entire system expands outward to engulf and kill surrounding trees.

Opposite: Pine trees are sometimes in the spring's way. If they are, hot water finds the trees, kills them and sometimes buries them in travertine.

Canary Springs is named after the yellow sulfur deposits, created when *Sulfurihydrogenibium yellowstonense* consumes dissolved sulfide and excretes elemental sulfur minerals. Successful terracettes grow upward and in the process entomb smaller upslope terracettes to form meter-high buttresses

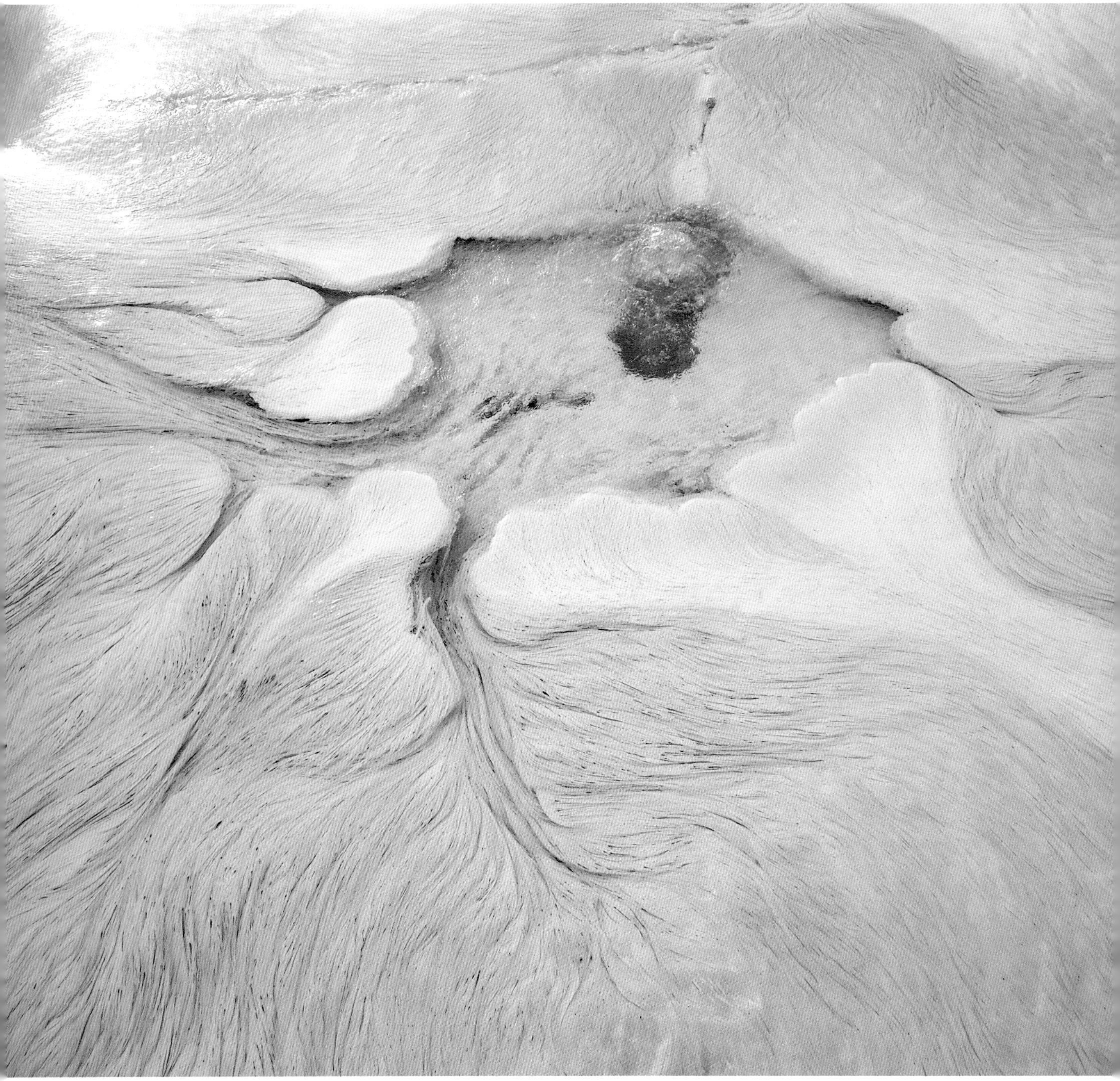

Opposite: Large bubbles of $CO_2$ effervesce as they emerge from 73°C turquoise water that emerges from a 10 cm diameter vent orifice. The hot-spring water quickly cools to 71°C, allowing light beige filamentous *Sulfurihydrogenibium yellowstonense* bacteria to form the fettuccine travertine streamer rock textures.

Left: Vent and Apron and Channel Facies deposits that form across geothermal gradients.

## INQUIRY IN SCIENCE

Scientists and artists love surprises. The unexpected opens the mind to learning, whets the appetite for new knowledge, and leads us into further and deeper investigation. The "scientific method" that every child learns about in grammar school begins with the construction of questions. These are the basis from which experiments and inquiry proceed. Unfortunately, we often make that process seem too linear and mechanical, a straightforward march from one step to the next, with too little recognition that one insightful and well-phrased question often spreads out to become many more. As a result, progress in the sciences is not so much a forward march as it is a spiral of inquiries and action, circling us closer to the truth, but rarely in a direct and easy way. This is the process we call ***scientific inquiry***.

intuition, and familiarity with other scientific work. We talk to colleagues, make estimates and develop assumptions. Stage Two in the process is to ***reason, plan and hypothesize***. Here we combine our findings and speculations into a viable experimental design that also blends art and science. This process requires reconnaissance, thought-experiments, and some rudimentary data collection from small-scale pilot experiments. We might do this work in the laboratory, out in the field, or conceptually on computers, or merely by thinking long and hard. Our goal is to establish the contextual framework we need to address these unanswered questions in ways that will ultimately bring results. For fieldwork in the natural world, this Stage requires time in the wilderness with your subject of interest, open-mindedness and an ability to integrate art and science with life experience and knowledge. Again, it is important to do this exploring with an open mind, to absorb and conceptualize what we did and did not expect to see on these journeys.

The Third Stage is to ***experiment***, collecting new observations, photographs, data and measurements to answer and test our targeted questions and hypotheses. Stage Four is to ***synthesize, interpret and discuss results***. This includes thorough integration and evaluation of the experimental results and incorporation of aspects of Stages One through Three. To do so means following strict protocols, such as instrument calibration and the collection of data in triplicate. Stage Five, when we are ready, is to ***communicate results*** via publications, lectures, workshops, classroom, field courses and other types of meetings to achieve formal and informal educational outreach. And finally, Stage Six is to ***modify, rethink and re-initiate curiosity***. The ultimate hope of art and science is always to lead us to new perceptions, by thinking and imagining, to ideas, questions and hypotheses that we could never have had before.

Opposite: Every late afternoon into dusk year round, bacteria may be able to conduct anoxygenic photosynthesis in hot oxygen-depleted water using salmon colored pigments. The salmon pigments fade away every evening after this daily process is completed.

Left: Ponds are encrusted with light beige to white filamentous microbial mats of the Apron and Channel Facies

Surface tension highlights the scalloped terracette margins. Lower temperature travertine features are growing in glassy smooth shallow pools.

## INQUIRY IN THE ARTS

Creative, open-minded inquiry in art, as in science, is sometimes boosted or even revolutionized by new technology. Much has been written, for example, about how the Impressionist painters in France benefited, with regard to the vitality and vividness of their work, from a manufacturing breakthrough in the middle of the 19th century. Oil paints became commercially available that were pre-mixed in metal tubes, allowing artists who worked in oils to do most or all of their landscape work in the open air with the scene right in front of them. Before this innovation, the actual color-mixing and brushwork had to be done back in the confines of a studio, where the light was different and the ephemeral moment of the day in the field was gone. Another example is the recent spectacular advancement in the resolution and overall quality of digital photography. Even so, the painter and the photographer must still master an array of prodigious skills, and practice with open-mindedness, intuition, creativity, timing, and a deep understanding of place.

Right: Image of the complex and immense Carina Nebula taken in 2010 by the NASA Hubble Space Telescope

Opposite: Narrow ridges of travertine only 0.5 cm in height form microterracettes that are perpendicular to the water motion. A diverse assemblage of microorganisms, shown by the light brown pigments, takes advantage of the accelerated water flow at the top of each razor sharp lip. Small white shrub-like clusters of aragonite grow within each ponded area.

Opposite: Rounded spalling patterns reveal older mounds beneath the last veneer of travertine. The gray coloration comes from microbes that have invaded the outer surface of dry travertine.

Left: A lone dead tree being engulfed and fossilized by rapidly growing travertine on Main Terrace.

In this way, and much like the scientist, the artist helps us to see the world with fresh eyes, to engage the natural context in original, free and daring ways. Art and science express the same basic human need to communicate, to seek, find and express the truth about the world and the human condition. Thomas Moran's watercolor sketches of the Yellowstone region in 1871, were completed on one of the first expeditions to survey and comprehend the wonders of the land that eventually became our first national park, while conveying the excitement and freshness of that early encounter. Helping to win support for saving the area, Moran's paintings also helped kindle scientific curiosity about the geology, dynamics and geothermal features of Yellowstone. Art and science take the long view, seeking to express and preserve truth for future generations.

Good photography in natural settings, the kind of photography that opens up the world to us, has other qualities in common with good science, among them a need for patience and for adaptability. Leaps in technology have not eliminated the need to go long distances in sturdy boots and with the right protection against challenging, changeable weather. Once you get where you think you need to be, you have to settle in, observe and wait. If the spot you have chosen does not prove to be right, does not open up new perceptions or possibilities, then you shift your ground and stay watchful again. Once you do have what you want, you are on to another quest, the next story from the universe of Yellowstone that you can help to unfold.

The gentle calm turquoise color of this pond is contained by scalloped edges. Previous ponds are now flooded by younger hotter rising water, supporting only microbes with little to no pigmentation.

Satellite image of the Mammoth Hot Springs complex in 2013. Actively flowing hot springs, represented by light to dark brown areas of microbial mat pigments, make up only a small proportion of the surface of Mammoth Hot Springs at any one time. The white is travertine in previously abandoned drainage patterns.

# DYNAMICS

## TRAVERTINE GROWTH, SHAPE, AND COMPOSITION

MAMMOTH HOT SPRINGS is created, sustained, and constantly transformed by water. Abundant and flowing from deep underground, it is the most ancient, precious and essential substance on our planet. All life depends on it, including the heat-loving (thermophile) microorganisms we find thriving here in temperatures and conditions where most other living things cannot survive. Heated by the Yellowstone supervolcano, these enormous volumes of hot water keep everything going at Mammoth. Less obvious, but equally important for the creation of this amazing and complex landscape, is the fact that the famous travertine "terraces" of Mammoth, the mineral deposits that give shape and grandeur to the springs, are directly precipitated from this same flowing hot water. Unlike common sandstone and shale deposits that result from the erosion and transport of sedimentary grains, the components of the Mammoth travertine are brought to this site in the form of chemical ions dissolved in the hot water. In this chapter, we will follow the remarkable impact that this phenomenon has had upon the geological, chemical, and biological dimensions of the Mammoth Hot Springs system.

The water we see emerging into the daylight at Mammoth is completing a long hydrologic journey that began as rain or snowfall in the Gallatin Range, seeps into deeply buried rock and rises back to Earth's surface in Yellowstone. The entire cycle can take from two thousand to eleven thousand years, during which time many notable changes happen to the water. The fresh rain and snowmelt chemically react at high temperatures and pressures with Madison Formation limestone thousands of meters underground, in a process called subsurface water-rock interaction. The water is transformed into a seawater-like fluid rich in calcium (Ca), magnesium (Mg), sulfur (S), sulfate ($SO_4$), sulfide ($H_2S$), carbon dioxide ($CO_2$) and other dissolved components. This change is no surprise, given that the Madison limestone was originally formed in shallow ocean environments about 350 Mya, thus retaining the chemical constituents of seawater in the animal and plant skeletons of which it is comprised. In the hot and acid-neutral spring waters emerging from Mammoth, these dissolved elements come together to bond with each other physically and chemically, precipitating the solid form of calcium carbonate ($CaCO_3$) limestone called travertine. The name travertine comes from Roman times in a famous region in Italy. For tens of thousands of years, hot springs in the Lazio region surrounding the ancient city of Tivoli have created thick deposits of *travertinus* (the ancient Roman name for Tivoli), a favored building material since the days of the Roman Empire. Travertine (the modern English and French version of the word), also sometimes called *carbonate sinter,* is often confused with coveted white *marble* like that mined in Carrera, Italy, and chosen by Michelangelo for his great sculptures. The term *marble,* however, refers to metamorphic rock, not sedimentary. Marble is in fact a limestone deposit that, over a long expanse of geological time, has been buried deep in the subsurface, undergone high-temperature chemical and physical alterations there (metamorphism), and turned into a denser, harder form of $CaCO_3$ that eventually rises during mountain building processes back to the surface.

This spring on Angel Terrace consistently flowed for a full decade and was the site at which the five-fold travertine facies were initially observed and described. The 1 m high front margin of the Pond Facies in the center was deposited in only one year.

The rapid growth rate, and the resulting immense weight of the travertine accumulating at Mammoth every day is astounding. Long-term field experimentation has determined that travertine here grows as much as 5 mm in a single day. From the perspective of geological time, this is almost unimaginable, being *one million to one billion times faster* than limestone depositional rates seen in other common Earth environments such as oceans, lakes, rivers and caves. The result at Mammoth is the deposition of a thousand kilo (a *ton*) of new travertine on a daily basis all across the entirety of the hot springs complex. The reason travertine has not covered the entire northern half of Yellowstone National Park is that travertine deposition occurs solely within the hot-spring outflow channels, which are relatively small and ephemeral, covering less than 10% of the total surface area of Mammoth at any one time.

These ever-shifting springs and the amazingly rapid growth of the travertine at Mammoth create a truly unique natural setting, one where the only constant is change. The landscape here is reshaped over and over again by three fundamental and evolving travertine manifestations (geomorphologies).

Right: The four-fold sequence of 50 m thick travertine terraces looking south from the hills behind the Mammoth Hot Springs Hotel. Liberty Cap dome is in the center foreground and Bunsen Peak rises in the back.

Opposite: Liberty Cap, near Palette Spring, is one of the most iconic landmarks at Mammoth, looking identical to its present form in the first black-and-white photographs of Mammoth in the 1870s.

The first are ***terracettes,*** the classic stepped limestone deposits of Mammoth, usually about 0.5 to 1 meter in both height and diameter. Simultaneously being formed are larger ***terraces*** (10 to 100 times the size of the terracettes) as well as smaller ***microterracettes*** (10 times smaller than the terracettes). All of these features retain the same stair-step shape, with a flat top and an inclined front margin. Few other features in nature exhibit this level of conformity in basic shape across such a range of sizes, an attribute that makes places at Mammoth like Canary Springs on the southern margin of Main Terrace so enthralling to observe.

Another basic travertine category is the ***dome,*** where hot water emerges from a single vent only a few centimeters in diameter. Unlike the water action that forms terraces, where water erupts from similar sized vents and flows downslope in a hemispherical pattern, the water in a dome flows out and away from the vent radially in all directions, creating meter-sized cylindrical accumulations of travertine. The most striking and long-lived examples of travertine dome formations are Liberty Cap and Devil's Thumb at Palette Springs. A ***fissure ridge***, a third travertine category, is created when hot water emanates from along the entire length of a fracture or fault. As a result, the travertine is deposited symmetrically along the two opposite paths of water flowing away from the fissure, forming the elongated ridges that we see at White Elephant Back.

Opposite: Devil's Thumb, the gray travertine dome in the middle right of the photograph at Palette Spring, has remained unchanged for more than 100 years. The new spring is revisiting this locale and threatens to bury the old formation.

Top left: Two travertine domes begin to form along a fissure ridge fracture with terracettes forming in the background. This illustrates how the basic shapes of travertine can be developed at a single spring system like Narrow Gauge.

Bottom left: Looking toward the Absaroka Range along the length of White Elephant Back fissure ridge.

The majestic Proximal-Slope Facies terracettes at Canary Springs, set against the gathering purple clouds of a storm over Gardiner.

Mammoth is a rare feature, sitting as a jewel on the back of the Rocky Mountains. Its source is snowfall and rainfall over the Gallatin Range, which flows down into the ground, is heated by the supervolcano, chemically reacts with 350 million year old Madison Limestone, reappears at the surface and creates a bizarre habitat for thermophiles at Trail Spring. It is the good "Perfect Storm" where rare and unique elements come together and produce this natural feature.

Opposite: The Vent Facies travertine is extremely hard and dense, forming mounded growths in the uppermost area of the photograph that are riddled with tiny holes caused by $CO_2$ release.

Top: Development of ridged networks in living mats composed of cyanobacteria and other microbes in the Pond Facies

Bottom: Aragonite needles in the Pond Facies grow as small branching clusters to form travertine "shrubs" that are covered with light brown mats composed of cyanobacteria and other microbes.

White calcite hopper crystals grow on the underside of surface tension in cooler water. Aragonite crystals grow as ridged networks on the warmer bottom. Gas lift-off structures come from pure oxygen bubbles created by microbial photosynthesis that pull the light green gooey microbial mat material to the surface. Calcite ice forms small umbrella-like crystal sheets that float on top of a still pool of water. The same process happens along the inside of the terracette edge.

Top: Calcite ice sheets formed at the cooler surface of the Pond Facies, then broke and sank to the bottom of the pool.

Bottom: Calcified bubbles at the downstream margin of the Pond Facies.

Opposite: The lipped margin of the Pond Facies has an irregular fluted shape, which marks the beginning of the cascade of microterracettes that comprise the Proximal-Slope Facies.

Bottom: Terracettes of the Proximal-Slope Facies at Palette Spring.

The travertine crystal growth rate is so rapid that calcite grows on the surface tension of the water itself (calcite ice sheets) and also on gas bubbles. This is a phenomenon never witnessed in our everyday lives. Travertine deposited in the ***Proximal-Slope Facies*** (35-50°C, pH 7.0-7.5) forms terracettes and microterracettes in a variety of shapes and sizes. However, unlike the Pond Facies that is laid down nearly flat on the top margin of the terrace, the Proximal-Slope Facies forms on the steeply inclined slope that is downstream and beyond the terrace margin (the major break in slope). As is true for the Pond Facies, travertine in the Proximal-Slope Facies is composed of dendritic aragonite, as well as calcite ice sheets and encrusted bubbles. A significant transition then takes place due to dropping temperature and flow rate, and increasing $CO_2$ escape, thus creating the gently-inclined ***Distal-Slope Facies*** (25-35°C, pH 7.5-8.0), which is a travertine composed solely of calcite crystals that form terracettes, microterracettes and solid millimeter-sized calcite balls ***(spherules)***. A distinct difference is easily seen between the terracettes of these last two facies. Viewed from above, the Proximal-Slope terracettes form arced hemispherical shapes with high steep margins that close laterally. Conversely, the Distal-Slope Facies form elongated terracettes with extremely low, rounded, and irregular margins that do not neatly close laterally.

Opposite: The low angle slope creates these gold colored travertine terracettes with a lacey pattern. Narrow, elongated and shallow (1 to 2 cm high) pools connect laterally. The low temperatures of less than 35°C sustain a highly diverse group of microbes. This is a chemical environment of calcite deposition, unlike aragonite in the higher temperature regions.

Left: Small solid balls *(spherules)* of calcite travertine form in a shallow terracette of the Distal-Slope Facies. These spheres grow in place around microbial clusters and are not the result of travertine deposition during rolling. Pine needles are also rapidly encrusted.

This section of Palette Spring exhibits a broad range in water temperature, flow rate, $CO_2$ gas release and pH, nutrient supply, travertine growth rate and other environmental factors. Typically at Mammoth Hot Springs, from the Vent to Apron and Channel Facies, the microbial communities exhibit DNA diversity that rivals that of coral reefs and rain forests.

## GEOCHEMISTRY OF THE HOT SPRING WATER AND TRAVERTINE: A PERFECT STORM

The water at Mammoth undergoes major changes in geochemical composition as it flows along and through the hot spring drainage system. These alterations are caused by the complex interplay of both non-biological *(abiotic)* and biological *(biotic)* processes, each with considerable impact on the travertine that is actively being deposited. Two of the most influential natural abiotic processes at Mammoth are the cooling of the water from 73°C to 25°C (or even cooler, depending on the season), as well as the liberation of dissolved $CO_2$ *(degassing)*. The more $CO_2$ that is dissolved in the water at the surface, the more acidic the water is. As more and more $CO_2$ degassing takes place during the downstream water flow, the water becomes gradually less acidic, precipitating ***aragonite travertine*** at higher temperatures and ***calcite travertine*** at lower temperatures. Biotic processes along the way also have a significant impact. For instance, photosynthesis by cyanobacteria, combining $CO_2$, $H_2O$ and sunlight, creating sugars and oxygen *(photoautotrophy)*, is manifest in the composition of the rock as it changes the chemical composition of the $CaCO_3$. In the end, this process also serves to remove $CO_2$, therefore helping to foster travertine formation just as $CO_2$ degassing does. This chemical change is preserved in the travertine $CaCO_3$ crystals and is preserved through geological time in ancient travertine deposits.

Opposite: Unbeknownst to this killdeer, surface tension makes water build up at the margin of microterracettes in the Distal-Slope Facies, which is quickly released over and over again to form roll waves.

Left: A forest of white feathered aragonite shrubs grow on the bottom of a high temperature pond. Crystals grow upstream into the oncoming current from the upper right. Hoar frost crystals also grow into the wind. The front of the pond margin exhibits scalloped microterracettes.

The changing flow regimes that occur along the hot-spring drainage channels also have a profound influence on the variations we see in water temperature and geochemistry. In the Vent Facies, the water flows quickly out of the vent and directly into a confined bowl-shaped area, where it begins the process of $CO_2$ degassing and travertine deposition. As a result, there is little variability in the temperature and acidity of the water that precipitates Vent Facies travertine. At the other end of the continuum, the water that forms the Pond Facies exhibit the greatest variability in temperature and acidity. Because this water spreads over a large surface area, it cools quickly and unevenly, undergoes evaporation, and fosters fast and extensive travertine precipitation. In water that creates the Distal-Slope Facies, the variations in temperature and acidity are less, because a great deal of $CO_2$ degassing and travertine deposition has already taken place upstream, and the water is therefore more homogeneous and less chemically reactive.

One way in which we track abiotic and biotic processes in both the water and travertine at Mammoth is with carbon-isotope and oxygen-isotope analyses. We use these isotopes because they play a key role in creating the $CaCO_3$ chemical structure of travertine, and they are important to both photosynthesis and degassing. ***Isotopes*** are atoms of an element that have an atypical mass, a nucleus containing the usual number of protons but a different number of neutrons. The high mass isotope of hydrogen in "heavy water" is a well-known example. Because isotopes will act differently from standard-form elements during biotic and abiotic chemical reactions, mathematical estimates can be made for the mass changes to be expected during a natural process. These estimates are then compared with actual measurements from the field, of the isotopic values of the water and travertine. For example: the process of $CO_2$ degassing can alter the presence of carbon-isotopes in the spring water, but not of oxygen-isotopes. Conversely, factors such as temperature change, water evaporation, and microbial photosynthesis can have a strong effect on the oxygen-isotopes, yet leave the carbon-isotopes essentially unchanged. To further distinguish biotic from abiotic processes, isotopic measurements of sulfur (S), strontium (Sr) and other elements have also been completed on the water and travertine at Mammoth. Taken together, these studies indicate that an intricate mixture of biotic and abiotic processes is continually at work throughout the hot-spring drainage systems, and that their relative influence varies in each successive downstream travertine facies within the outflow drainage systems.

Filamentous microbial mats project just below the surface into a pond, as well as flow down the outflow. These two processes will soon join to completely cover and obscure the pond. A raft of white calcite ice, visible here in the lower right of the photograph, is supported on the top of the pond by surface tension.

Another hallmark of the hot-spring drainage systems at Mammoth is the large and rapid changes that take place over very short distances in the geochemical reactivity of the water and channel floor environments. These are measured according to their reduction-oxidation (redox) potential, which are the geochemical controls that dictate the way in which microbial life utilizes the natural energy resources of the springs. Reductants (reducing agents like $H_2S$) are chemical compounds that easily lose their electrons to other elements, a process called oxidation. Electrons lost or donated are then received by oxidants (oxidizing agents like $O_2$), a process called reduction. For any oxidation

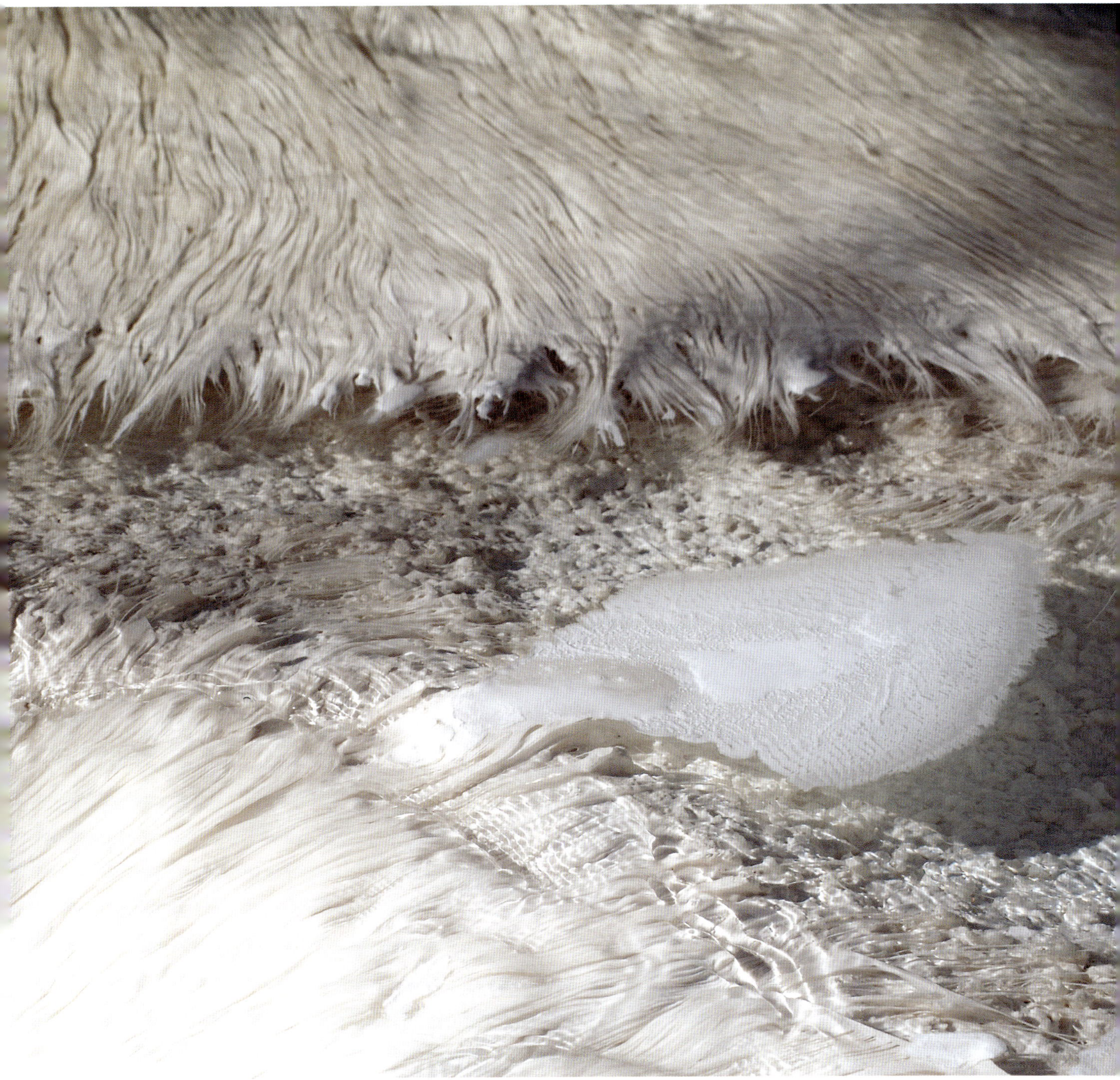

reaction to occur, a coupled reduction reaction must also take place because electrons will not float freely in water. These reactions result from biotic as well as abiotic processes, and strongly influence the acidity of hot-spring water via chemical transfers that can include whole hydrogen atoms (electrons and protons).

Especially important at Mammoth is the simultaneous redox cycling of hydrogen (H), oxygen (O), nitrogen (N) and sulfur (S). The latter is especially complicated due to the involvement of multiple S species and compounds. Cycling of S is also quite noticeable and sometimes even nauseating because of the ubiquitous rotten-egg smell of volatile and deadly $H_2S$ gases that the human nose can detect at only trace amounts (parts per billion levels). The source of the $H_2S$ is predominantly from the supervolcano, but some proportion might also be derived from the reduction of $SO_4$ in the subsurface Madison Formation limestone. The Apron and Channel Facies boasts the most extreme changes in redox potential. Here, over a downstream flow distance of only one to two meters, the water drops from 71 to 68°C and a pH of 6.2 to 6.4. They are also transformed from having high $H_2S$ and very low $O_2$, to being high $O_2$ with no $H_2S$. This "Perfect Storm" of steep environmental redox gradients was common at the dawn of life on early Earth, served as an environmental cradle that fostered the evolution and adaptation of ancient microbial life, and is still active right in front of us at Mammoth.

Opposite: Indian rice grass, a common plant in the area, often breaks off and falls into the pools. Immediately they become encrusted with aragonite crystals, the same process that forms the pond cotton balls visible on the bottom of this pool, until the grass is completely covered and vanishes. This water is supersaturated with respect to calcium carbonate ($CaCO_3$). The ions of $Ca^{2+}$ and $CO_3^{2-}$ bond because of their high concentrations in the spring water, and they therefore grow and quickly cover (armor) virtually any surface.

Left: Travertine deposition from higher water outflows create rounded (not sharp) pool edges. Light beige filamentous microbial mats grow on the upstream margin of the pools, where dark rust colored microbial communities live within the pools and in thin dark channels on the margins.

Where there was once high water flow creating classic stalactite-style formations, now low water flow is just enough to maintain the growth of thin colorful microbial mats on top of the older travertine.

## DIVERSE ANCIENT MICROBIAL LIFE

Thermophilic microbes at Mammoth Hot Springs live in thin slimy layers of cells and mucus on the bottom of the spring's drainage channels or just below the uppermost water surface (both are called *microbial mats*). Cells can also float in the water column (called *bacterioplankton*) and become airborne in steam (called *bioaerosols*). We have come to know the diversity and varied behaviors of these microorganisms from first studying their distribution, mat shapes and diverse pigmented colors in the field at Mammoth, then following up with molecular sequencing and microscope imaging technologies in the laboratory. These integrated studies have provided new insights into how microbial life at Mammoth has evolved and adapted to the high temperature waters and astounding rates of travertine crystallization. Importantly, this has proven that the hot springs at Mammoth are teaming with microbial life and thus are not the sterile barren environments they were presumed to be more than half a century ago. Only 100 m or so in the downstream flow from the Vent to the Distal-Slope Facies, the spectrum of genetic diversity at Mammoth rivals that of rain forests and coral reefs. All three of the fundamental types (domains) of life on planet Earth (Bacteria, Archaea and Eucarya) are present as part of the Mammoth Hot Spring ecosystem. Of these, only Bacteria and Archaea actually live within the flowing waters, while the Eucarya are represented by the menagerie of animals that regularly visit and utilize the natural resources of the springs.

Below: Shadows highlight the textures on the surface of microterracettes on a low angle slope.

Opposite: Aerial view of Grand Prismatic Spring and Excelsior Geyser Crater in the Midway Geyser Basin. The saturated dark blue pools look similar. However, Excelsior Geyser Crater runs hotter, flows faster, and is hydrologically focused into only one channel. This prevents significant growth of thermophilic microbial mats. Grand Prismatic Spring is cooler and outflow is distributed around the circumference, permitting extensive growth of brightly pigmented microbes in the myriad of channels. The people walking on the boardwalks near these features provide a measure of the immense size of the vents, which are three orders of magnitude larger than the vents at Mammoth.

Right: What looks like the texture of a kitchen sponge is actually hard drying travertine riddled with ridged networks. The field of view is 20 cm across.

Opposite: Recently abandoned pond travertine cotton ball accumulations retain their brown pigmentation. Hotter water flow had begun to deposit filamentous microbial formations across the surface of the cotton balls just before all the flow had completely stopped.

The prominent green pigments in the mid-temperature Pond and Proximal-Slope Facies are formed by cyanobacteria, no algae live in the waters of Mammoth. This reality is in stark contrast to the green thermophilic algae *Cyanidium* that thrives in the hot-spring drainage channels in places like the Norris Geyser Basin in central Yellowstone.

Microbial diversity at Mammoth increases dramatically downstream from high-to low-temperature spring water environments. Bacteria create clones to reproduce without sexual interaction. Therefore, it is difficult to use the word "species" when trying to distinguish different types of microorganisms, given that the term normally refers to creatures that are able to reproduce sexually. Instead, molecular sequencing now permits us to classify different types of microbes (most of which still cannot be cultured in the laboratory) according to measured differences in their genetic composition. These are referred to as *Operational Taxonomic Units* (OTUs) rather than as species. The 16S rRNA gene sequence, established by Carl R. Woese at the University of Illinois Urbana-Champaign (see Chapter 4), is the universally accepted gold standard for making microbial OTU identifications. Analyses targeting microbial mat and planktonic bacteria have identified 7 OTUs in the Vent Facies, 17 OTUs in the Apron and Channel Facies, 92 OTUs in the Pond Facies, 63 OTUs in the Proximal-Slope Facies, and 18 OTUs in the Distal-Slope Facies. In addition, three Archaea OTUs were detected in the Vent Facies.

It is important to recognize that these are conservative estimates based on multiple rounds of sampling and thus the actual biodiversity is expected to be even higher. In addition, the steam liberated from each facies carries a small grouping of the resident microbes up and into the air directly.

While microbial life is abundant and diverse at Mammoth, each OTU requires a specific range in temperature, acidity, flow, nutrient availability and other key conditions (its preferred *niche*). It was also unexpected that the microbial community living within each travertine facies is dramatically different from each of the other facies. Given that the water is continuously flowing through these hot-spring drainage systems, one would assume that microbes would commonly be washed from upstream to downstream environments. This happens, in fact, with two of the most thermophilic bacteria at Mammoth that are related to the *Aquificales* and *Beta-Proteobacteria*. These turn up as the only two groups of bacteria detected in all five of the travertine facies. It is surprising that they can withstand such a wide range of temperatures.The correlation of microbial communities with travertine facies in the springs at Mammoth today means that fossilized travertine facies provide a geological fingerprint, from which microbial diversity in ancient hot springs can be reconstructed.

But these facies-distinct distributions of microbial communities at Mammoth do not help us answer the vexing question of where these microbes originally came from, and how they ended up in each downstream position. Two Dutch scientists, Martinus W. Beijerinck and Lourens G. M. Baas Becking, from the influential Delft, Leiden and Wageningen schools of microbiology, have proposed an answer to these concepts: that "everything is everywhere, but the environment selects." This elegantly concise yet somewhat enigmatic hypothesis is being tested around the world in natural environments like Mammoth. Where do the bacterioplankton in the vent waters come from? Do they originate somewhere on the long hydrologic journey of the water from the Gallatins into the subsurface Madison Limestone and ultimately in the spring water eruptions at Mammoth? Or, since it is likely impossible for microorganisms to survive the high temperatures and pressures deep in the subsurface of the Earth's crust, are these microbes somehow introduced when the waters have reached the shallower and cooler subsurface sequence of fossil travertine at Mammoth? Or can microbial life, dormant in the form of spores, be transported so far and long without being active and instantaneously resuscitated only when the right environmental conditions become available? And how is it possible, given the continuously and rapidly flowing water in the hot-spring drainage channels, that these microbial communities are partitioned from one another within facies along the way, as though cleaved by an invisible knife? These are among the many unsolved mysteries for future generations of geobiologists to explore at Mammoth Hot Springs.

A dense miniature forest of 2 to 3 cm high crystalline aragonite shrubs formed on the bottom of a mid-temperature pond. Because it was recently drained, the white shrub tips of the microbial mats are drying and dying.

Delicate sweeping hair-like filaments brush across the cotton ball formations on the floor of the pond. Only the tips of these filaments are naked threads of the bacterium *Sulfurihydrogenibium yellowstonense,* which is a sulfur-consuming organism. Within hours, travertine coats these filaments to firm travertine crusts at the water's surface.

## MAMMOTH: UNEXPECTED PRODUCTS OF LIFE-EARTH INTERACTION

Abiotic and biotic processes are constantly waging an unseen tug-of-war to create the splendor observed at Mammoth. These struggles include physical and chemical attributes of the spring water such as temperature, acidity, chemistry, gas release, flow rate, and steam liberation, as well as the influence of all living components that include microbes, surrounding plants and visiting elk, bison, bears, birds, insects and humans. By approaching Mammoth with a combination of holistic and reductionist scientific approaches that integrate the key physical, chemical and biological components of the hot springs, a series of fundamentally new insights has been forged about how and why Mammoth came to exist.

### The Life History of a Hot Spring

The speed at which travertine is deposited in the hot-spring drainage channels at Mammoth might lead one to assume that the frequent closure and movement of the vents is due to travertine clogging the subsurface plumbing. Actually, the waters that erupt at Mammoth, supercharged with dissolved $CO_2$, are slightly acidic, making them geochemically incapable of self-sealing with travertine. It is not until the waters come into contact with the atmosphere and begin to liberate their burden of dissolved $CO_2$ that they are capable of rapidly precipitating travertine. This is why the vent openings at Mammoth occur at the bottom of shallow bowl-shaped depressions within the Vent Facies, instead of erupting from a travertine mound. Scientists postulate that these frequent shifts in the position of vents at Mammoth are instead due to the massive weight of the travertine itself. The natural sub-surface plumbing of Mammoth is composed of intricate branching networks of small straw-like conduits, which is in contrast to the type of large underground chambers that feed geysers. Gravity-driven compression slowly crushes these conduits as the travertine continues to build, eventually turning off the water flow.

The Vent Facies has a bowl-shaped depression because the hot turquoise waters are slightly acidic, preventing travertine deposition until it reaches the surface and liberates dissolved $CO_2$.

These dynamics create the drama of change and recurrence at the hot spring vents of Mammoth, which average five to ten years in duration. Over the decades, many new springs have suddenly gurgled to the surface, sometimes beginning their flow literally under our feet as we were conducting fieldwork. These eruptions have allowed us to track the life cycle of these springs from their birth through their gradual demise. The process begins when cooler (about 45°C) water suddenly reaches the ground surface in small 0.5 to 1 centimeter-diameter vents. Within two weeks, the flow increases, and the vents expand to two or more centimeters in diameter, at which point the erupting water will have reached the standard 73°C average temperature for surface flow at Mammoth. Over the next several years, the five-fold travertine facies will come into being along the outflow drainage system. Then, for reasons that remain unknown, the volume of water exiting the vent will dramatically increase, and remain at this increased flow rate for several months. This "midlife crisis" in the lifespan of a spring is then followed by a slow, continuous decline in the volume of water emerging from the vents and fissures,

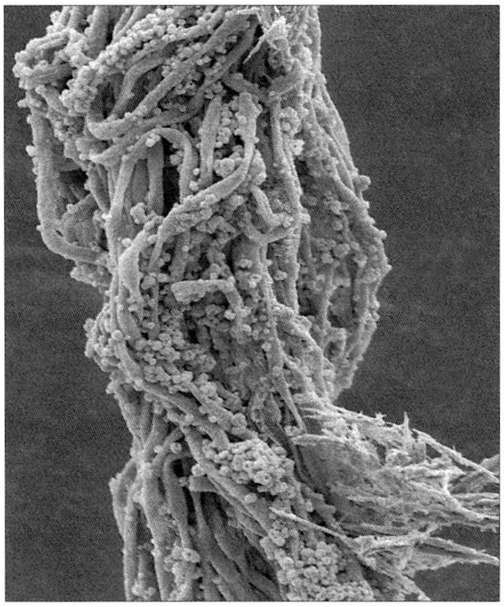

Top left: SEM image reveals that the long *Sulfurihydrogenibium filamentous* strands (10 micrometers thick in this sample), are composed of individual cells that intertwine with each other in the hot and turbulent waters of the Apron and Channel Facies.

Top right: SEM image of small (1 to 3 micrometer long) densely packed aragonite needles encrusting the surface of *Sulfurihydrogenibium* filaments, while larger (10 micrometer-long) and less densely packed aragonite needles grow on the EPS mucus draped in the spaces between filaments.

Opposite: Naked cappellini *Sulfurihydrogenibium yellowstonense* filaments project into the downstream pool. The question remains whether or not each microbial filament remains viable and active as it becomes more and more encrusted with travertine in the upstream direction.

The downstream margin of both the fettuccine and cappellini regions are composed of naked microbial filaments that wave like loose streaming hair just below the surface of the current and grow laterally at amazingly rapid rates that can reach centimeters per day. High-resolution microscope imaging and x-ray computer tomography show that the microbial filaments, each of which can reach tens of centimeters in length, are composed of individual intertwined microbial cells. Aragonite needles encrust these filaments, rapidly forming travertine streamers that become thicker and more solidified in upstream positions. The aragonite needles growing directly on the microbial cells are small and densely packed, while the aragonite needles growing on EPS mucus in the open spaces around each cell grow large and are loosely packed. Remarkably, when not in the rapidly flowing water of the Apron and Channel Facies, the individual small ***Sulfurihydrogenibium yellowstonense*** cells detach from one another to form micrometer-sized jellybean shapes and no long filaments. Because of the rapid travertine entombment process, all detailed aspects of the complex textures and fabrics of the fettuccine and cappellini can be found perfectly preserved and fossilized in ancient travertine.

When the volume of spring water emerging from the vents dramatically increases during the midlife crisis of the springs, it causes hot water to push farther downstream. This change results in Vent and Apron and Channel Facies being deposited in new downstream positions, over the top of lower-temperature Pond, Proximal-Slope and Distal-Slope Facies. Bypass channels, which look bleached because the white travertine can be seen through the transparent high temperature microbial cells, move water down and through the downslope lower temperature facies. This results in travertine overprinting, where higher temperature facies are deposited downslope in lower temperature facies. These exquisite and complex layering patterns of travertine and their bypass channels are well expressed at Canary Springs.

There are surprising similarities between the shape of the fingers of falling water and the tooth-like travertine encrusting microbial filaments. The large overhanging ledge at the top forms because the water moving at high velocity shoots over the edge and the microbial filaments and travertine follow the water out and down. The increased water flow creates these features at the top and is overflowing and burying the older slope terracette deposits below.

The most dynamic response to the stronger and hotter water flow is exhibited by the *Sulfurihydrogenibium yellowstonense* microbial mats of the Apron and Channel Facies, which accelerate their horizontal growth out and over the immediately downstream Pond Facies. Because these *Sulfurihydrogenibium yellowstonense* filaments growing just below the surface of the water are quickly encrusted to form travertine streamers, they can eventually cover an entire pool surface with a rigid lid of streamer travertine. The hot spring water can rush across the top surface of this streamer pavement and pour over the margin of the terracette. The *Sulfurihydrogenibium yellowstonense* filaments then follow the hot water, growing over and down the pond-lip margin to form vertical stalactite tooth-like cones of travertine. When the flow is strong enough, the hot water can sometimes fountain horizontally off the pond margin, allowing the *Sulfurihydrogenibium yellowstonense* filaments and ensuing streamer travertine to build out into thin air.

Orange Mound Spring is a fissure ridge, which means that low volumes of water flow out from multiple small vents along a narrow fracture line. This results in slow travertine deposition next to only the fracture itself, causing the feature to grow straight up.

This distinct pattern of Apron and Channel Facies streamer travertine, covering and hanging over the edge of the underlying Pond Facies, is the travertine fabric most commonly and easily observed from the boardwalks and roadways at Mammoth. It is also a travertine layering pattern that can be created only by the entombment of filamentous microbes, thus creating another reliable biomarker to search for in ancient travertine deposits.

Another important example of biotic control on travertine deposition, herein called the "Gas Lift-Off Sequence," occurs in the Pond Facies. Communities of brown and green photosynthetic cyanobacteria, such as *Synechococcus* and *Spirulina*, form thin flat-lying mats on the floor of the ponded pools. As these microbial mats continue to grow and develop, the active photosynthesis within the mats generates pure $O_2$ gas. Because the mats house active microbial cells and abundant gooey and stretchy EPS mucus, the $O_2$ becomes trapped in the mats instead of being released to the overlying water column. As these small gas bubbles accumulate into larger ones, they generate enough buoyant lift force to raise portions of the microbial mat up into the water column. The first stages of this process are seen as "ridged networks," where ridges caused by tension in the rubbery mats run down in irregular patterns away from the central bubble. This is a morphology similar to what we see in a circus tent, where a central pole, lifting and stretching the canvas, creates ridgelines. The end result is something that might suggest "alien" brain matter in a science fiction movie. This process can then move to the next stage, a "Gas Lift-Off Column," where the $O_2$ has accumulated into much larger gas bubbles and the cyanobacterial mats are less rubbery. At this stage, the bubble generates a strong vertical lift force stretching the mats into a cylindrical tower. Eventually, this column will reach the water surface of the pond, where a layer of calcite ice sheets begins to grow in concentric circles, radiating from the central gas bubble. All of these shapes and forms in the cyanobacteria mats become encrusted within the travertine deposit and preserved as biomarkers for our research into ancient dynamics.

Opposite: A latticework of aragonite formed on the bottom of a recently drained low temperature pond looks like a mat of vegetation, but it is not. The lines of the ridged network formed when photosynthesis-derived oxygen bubbles pulled microbial mats upward. The creases of these distorted mats form the ridge network. The place where these ridges join is the site of the original oxygen bubble.

Yet another important category of biomineralization at Mammoth involves the production of specific biomolecules, such as proteins, polysaccharides, lipids and EPS mucus that microbes form in their outer cell walls as a natural product of their daily metabolic activity. By conducting controlled experiments with the Apron and Channel Facies, both in the field at Mammoth and back in our laboratories, we have proven that these membrane-bound molecules in ***Sulfurihydrogenibium yellowstonense*** dramatically accelerate the rate at which travertine can grow. Without the microbial mats, travertine will continue to form, but much more slowly and with distinctly different shapes. This natural phenomenon, also evident in the shell growth of clams and other marine animals, may even influence biomineralization such as kidney stone formation within the human body.

One of our original hypotheses two decades ago, when we were starting our work at Mammoth, was to determine whether or not the stair-stepped travertine deposits (terraces and terracettes) were biomarkers

Increased flow is burying and creating a whole new landscape on the slope of Minerva Terrace. The original older terracette structure, still visible in the upper left, is being buried in the center and right under striated stalactite-like fingers and porcelain smooth coatings.

## Fossil Record of Travertine

Another vitally important reason that we chose to work in this northern region of Yellowstone is that a thick sequence of fossilized travertine was deposited nearby from ancient hot-spring environments identical to those in the active springs at Mammoth. As a result, our research into the water, microbe and mineral interactions controlling modern day travertine deposition here can be directly applied and tested on these fossil spring systems close at hand. Uranium-isotope dating indicates that the 73 meters of travertine deposits underlying Mammoth, spreading atop rhyolites of the Huckleberry Lava Flow, began 7,700 years ago and have continued through the Holocene Epoch and into the modern day. Conversely, travertine deposits laid down near the town of Gardiner, a series more than 100 meters thick, formed from about 19,500 to 38,700 years ago, during the Pleistocene Epoch. These lie atop basaltic lava flows and are overlain by Pinedale glacial deposits.

The United States Geological Survey (USGS) drilled the Y-10 core at Mammoth in 1969. This effort provided travertine samples for dating, as well as for chemical and microscope analyses. Analyses of these samples, as well as old travertine at the surface of Mammoth, tell us that as you go deeper and deeper down into older Mammoth, the chemical alteration of the travertine increases. All chemical, physical and biological alteration of a travertine after it was originally deposited is categorized as ***diagenesis***. Around Gardiner, as the Pinedale glacier retreated at the start of the Holocene, large volumes of fresh meteoric waters flowed through the local travertine deposits, transforming them into extremely hard and well-cemented limestone, which has been commercially quarried from time to time over the past few decades. In many places, this water-rock interaction process served to dissolve the original travertine and precipitate

Opposite: Pinedale glacier tills (gray) deposited on top of a series of 30,000 year old travertine deposits in the Gardiner Quarry.

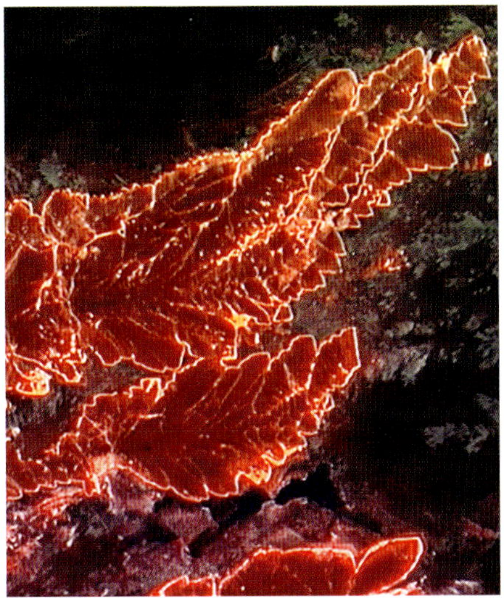

Bottom left: Microscope image of a paper-thin slice of travertine, collected from New Highland Terrace, which is being bombarded with electrons in a vacuum chamber. Chemical elements comprising the travertine, in this case manganese (Mn), emit bright cathodoluminescence light (CL) under these conditions that show geometric patterns of crystal growth. Bright orange calcite crystals are growing into a field of dark aragonite crystals in the Proximal-Slope Facies.

Bottom right: A microscope CL image of Pleistocene-age Pond Facies travertine collected from the Gardiner Quarry. Multiple geometric growth patterns of calcite exhibit Halloween-like brown and orange colors that were formed during chemical alteration of the travertine *(diagenesis)* that likely took place during melting of the Pinedale glacier.

Top: This smooth wall of 30,000-year-old Gardiner travertine was cut by expert quarriers brought over from the Tivoli region of Italy. Irregular dark lines preserve the cross section shape of the original terracettes and microterracettes that formed in this ancient Proximal-Slope Facies deposit.

Bottom: Apron and Channel Facies streamer travertine deposited from Pleistocene hot springs in the Denizli Basin of Turkey.

new $CaCO_3$ minerals. However, we also find original large-scale rock fabrics and some chemistries that have been well preserved and fossilized, allowing identification of the volume and distribution of the five travertine depositional facies within the Gardiner quarries. Large clean-cut quarry walls also show us a detailed history of how the travertine can simultaneously accumulate in both vertical *(aggrading)* and horizontal *(prograding)* directions to make an overall three-dimensional deposit of ancient travertine. It is common to find here, as well as in other quarries around the world, excellent samples of Apron and Channel Facies streamers, Proximal-Slope Facies microterracettes, as well as Distal-Slope Facies microterracettes, spherules and encrusted plants from well before the last Ice Age.

Above: Fossilized travertine microterracettes of Proximal-Slope Facies in the Gardiner Quarry.

Left: Perfectly fossilized plants, possibly sage brush, within microterraces that are in 30,000 year old travertine from the Gardiner Quarry. The vertical lines are drill holes from the quarrying process.

Proximal-Slope terracettes near Minerva showing diversity of microbial pigmentation.

Bison cow, enduring an early morning -37°C temperature, is coated with frost.

# EMERGENCE OF LIFE

*You have to have your own particular sensitivity to the world and there has to be parts of it that are beautiful simply because they are beautiful to you, regardless of what anybody else ever thinks. You see this all the time in an artist and you see it also in good scientists.*

— CARL RICHARD WOESE, 1928-2012

## A FAMILY TREE

TO BEGIN TO UNDERSTAND the organisms living in and around Mammoth Hot Springs, we have looked at them as part of the structure and history of all life on our planet. We have gazed into the Mammoth ecosystem as a window on the universe back to the farthest reaches of time. It is often observed that life on Earth is composed of "stardust" that came into being with the Big Bang many billions of years ago. These celestial raw materials, relentlessly transformed and reshaped, provide molecules that are the building blocks for the breathtaking diversity of life. This staggering display at Mammoth and on our planet now is but a small sampling of the trillions upon trillions of organisms that have taken form, flourished, and gone extinct during the long geological history of the Earth.

Opposite: This pudding at Fountain Paint Pots is formed from heated meteoric waters that are so acidic they dissolve the bedrock. New clay minerals and microbes grow in the slush. The acidic steam and bubbling clays are thrown into the air and form the cone.

Left: Bald eagles soar easily with broad rounded wings. This efficient way of traveling allows them to hang in the sky watching for prey for long periods of time with little effort. Their eyesight is remarkable. What we can see clearly at 20 feet away with 20/20 vision, they can see as clearly at 100 feet away. We see three basic colors, they can see five basic colors including ultraviolet light which helps them to see UV reflecting urine trails of small rodents. The light detecting cells, called cones, on the backs of their eyeballs are much more densely packed, so they can resolve five times more detail than us.

How have we come to know so much about the evolution of the biological world around us, and to understand the relationship of that long history to modern biodiversity? Living in the narrow confines of the here and now, we face a real challenge in imagining the work of evolution over vast expanses of time and the basic truths and dynamics that have made so much difference in the course of the planet's history. There are facts here that should matter to everyone, all over the world, as we participate in decisions about our collective future. The origins and development of microbial life need to be understood widely if we are to make headway in coping with the environmental crises that the planet faces now: to generate clean energy, to improve health care, and to explore for life on other planets.

To get our bearings, let us begin with something familiar and relevant—the idea of a family tree. Anyone can sketch a few branches of their own family and know what they mean: your grandparents gave birth to your parents, who gave birth to you, and so on. Everyone on every branch in your tree is descended from or connected to everyone else. Similar diagrams of lineage and descent are pervasive in the biological sciences, where the objective is to understand relationships between species and to gain insight into the shape and structure of the ultimate family tree, the tree of all life, past and present, on Earth.

Now you may wonder how we could possibly construct such a family tree of all life, going back to its very beginnings. Amazingly, we actually *do* know much of it now, based on the revolutionary work begun more than fifty years ago by Carl R. Woese, a Professor in the Department of Microbiology at the University of Illinois at Urbana-Champaign. His research literally redesigned and newly envisioned the family tree of life on this planet.

## EVOLUTION OF THE TREE OF LIFE

Keeping track of genealogy is an ancient practice and for the sake of wealth, power, and social position, people have been concocting family trees for thousands of years. In the biological sciences, however, there was no serious effort to adapt the methodology until the end of the 17th century, when the world of microbial life finally became open to human study. In 1674, Antonie van Leeuwenhoek, a textile merchant and self-taught scientist in the city of Delft in the Netherlands, constructed the first working microscope. Composed of a small brass plate, a handmade spherical glass lens, and a needle-like point, positioned with set screws, to hold samples for study, this simple looking device was revolutionary in its moment, and with it van Leewenhoek became the father of microbiology. Making the first observations of microscopic life in lake water, he published his findings in the Dutch *Proceedings of the Royal Academy of Science.* Today, one of the most prestigious awards in microbiology is the Leeuwenhoek Medal—an honor bestowed on Carl Woese in 1992.

In 1735, about half a century after Leeuwenhoek began his studies, Carl Linnaeus, a Swedish scientist also known as Carl von Linné, established a systematic approach (*taxonomy*) for describing and classifying living organisms, publishing it in a document he called *Systema Naturae.* As he revised and developed the system during his education at the University of Leiden in The Netherlands, and afterwards as a professor at Uppsala University in Sweden, his *Systema Naturae* grew from a pamphlet into a multi-volume treatise. Linnaeus' taxonomy organized all living things into only two groups or "kingdoms," *Animalia* and *Vegetabile.* His third proposed "kingdom," *Lapideum,* encompassed the inanimate world of rocks, minerals, and fossils. Linnaeus' taxonomy, organizing plants and animals according to shared characteristics, paid scant attention to environmental conditions in which a given organism lives. This kind of thinking broke from the conventional context for taxonomy used in scientific work at that time, and long after. Linnaeus originally concentrated his attention and classification on reproductive organs, later expanding to include all morphological characteristics. He clarified and simplified our thinking about the biosphere, giving us two basic distinctions that remain fundamental in our thinking today: *genus* and *species*.

Photograph of Carl R. Woese at the light table on which he painstakingly analyzed molecular patterns of microbial ribosomes that were radioactively exposed on x-ray film. These tedious and methodical measurements led to the identification of the 16S rRNA gene, which is the fundamental yardstick for understanding the evolutionary relatedness of life.

The middle of the 19th century saw explosive developments in the quest to understand and map out the history of life. Alfred Russel Wallace and Charles Darwin were the first scientists to introduce the concepts of evolution and natural selection as serious and formidable contributions to our understanding of the history and dynamics of life. Darwin's *Origin of Species* (1859), the first study to link the branching patterns of heredity with evolutionary processes, recommended that taxonomic classifications be based henceforth on these evolutionary relationships. However, because *Origin of Species* focused exclusively on animals and plants, it did not include insights about single-celled microorganisms. It was not until 1866, in work by Ernst Haeckel (a Professor at the University of Jena, Germany), that microbes were first included in a tree of life.

Pine, fir and cottonwood trees along Pacific Creek in Grand Teton National Park in September. These mountains have changed dramatically through time, experiencing kilometers of uplift over millions of years. This view was sculpted by many previous glacial events, the last one was the Pinedale Glaciation, 15,000 years ago when at least 1.5 km of ice accumulated here.

## MOLECULAR PHYLOGENY

Prior to Woese's work, it was thought that there were only two general categories of life on Earth—Eukaryotes and Prokaryotes. The single-celled (microbe) prokaryotes were conventionally described based on only two criteria: the biochemical (metabolic) capabilities they show when grown in laboratory conditions (*as isolates or pure cultures*) and their shapes and structures seen with a microscope. Though culturing can tell us much about how a microorganism functions, it is still estimated that fewer than 1% of the microbes on Earth can be cultured in the lab. Moreover, even if we do succeed in culturing a given microbe, the conditions under which it thrives in nature, including the complete community of life with which it interacts and on which it may depend, can only rarely be reproduced in the lab.

Two grizzly bears courting in early June on Mt. Washburn.

Field of Great Basin wildrye grass with innumerable numbers of stems and dramatic sunset lighting.

Because microbes can be so hard to distinguish from one another—as spheres, rods or filaments with similar internal structure—even a careful study of their biochemical and morphological attributes may not provide a crisp, unambiguous determination of their evolutionary relatedness. The problem is a bit like relying on traits such as hair color and height to determine kinship among human beings; these markers can tell us something, but not enough to give us reliable and precise indications of family ties.

Carl Woese, therefore, began his quest with a hypothesis that in order to create an accurate universal tree of life, some trustworthy indicator of evolutionary relatedness had to be found in the cells of every living creature. Born in Syracuse, New York, in 1928, Woese studied mathematics and physics at Amherst College and completed his PhD in biophysics at Yale in 1953, under the direction of Professor Earnest Pollard. After a few years at the General Electric Research Laboratory, Woese joined the University of Illinois as a tenured professor in Microbiology, an unprecedented honor supported by colleagues there, including Professor Sol Spiegelman. Arriving in Urbana as a relative outsider to the ways and mind-sets of classical biology, Woese had no preconceptions about how evolution is "supposed" to unfold. He did not set out to overturn Linnean classification and Darwin's theories of evolution; he simply wanted taxonomy to reveal and reflect evolutionary relatedness.

The wildflowers in this photograph include buckwheat, lupine, geranium, cinquefoil and sage. This colorful wildflower display, while impressive, represents only a dozen or so plants. Whereas a similar color array in the hot-springs would represent many hundreds to thousands of times more biodiversity.

The word ***phylogeny*** is derived from the Greek roots ***phylon,*** meaning stem, and ***genesis,*** meaning origin. ***Molecular phylogeny,*** as Carl Woese developed and practiced it, establishes evolutionary relatedness by measuring and comparing the chemical composition of key biomolecules within cells. The ***genetic code*** is the combination of nitrogen-containing nucleotide base pairs (A-T Adenine-Thymine, G-C Guanine-Cytosine), to create alternating sugar-phosphate chains corresponding to specific amino acids—the building blocks of protein. One way to envision the A, G, C, and T molecules is that they are the musical notes of life. In Western music, it is hard to imagine that all of the diversity of melodies is composed of different combinations of just eight basic notes on a musical scale, which are arranged and rearranged in different combinations to make unique scores for every song. Similarly, in DNA, the four basic nucleotide base pairs have been arranged and rearranged to make a unique score for every living organism. Woese believed a universal tree of life was possible because the genetic code, the score of life, is consistent in all life forms. Humans are created from the same genetic code as

Yellowstone bison, which is also the same genetic code as bears, moose, eagles, rattlesnakes, beetles, lodgepole pines, rye grass and every organism in the Park. The genetic code turns out to be extremely good at minimizing errors. If a permanent alteration is made in the nucleotide sequence of an organism's genome (a *mutation*), the resulting amino acids will closely resemble the ones in the original form. In other words, although the genetic code has evolved over time, its robust stability and overall consistency permit development of an accurate molecular phylogeny.

In 1964, just a few years after joining the faculty at Illinois, Woese began an ambitious program to construct a molecular phylogeny of all life. At that time much of the attention in the life sciences was focused on furthering the work of the American biologist James Watson and the English physicist Francis Crick, who had shared the 1962 Nobel Prize for physiology and medicine for their description of the double-helical structure of DNA and the genetic code. Working from an insight that knowing biological *structure* was not sufficient for understanding biological *function*, Woese wrote to Crick about his ideas, but Crick's response was not encouraging. In his view, Woese's overall objective was probably impossible to attain; the proposed work would likely produce only limited results, and securing financial support for it would be difficult. Crick's reaction was typical of how the mainstream scientific community would react to Woese's radical thinking for years to come. But Carl Woese was undeterred.

Left: Garter snake, one of only six reptiles in the Park, relatively common but shy.

Cells must reproduce to make future generations, they must metabolize by extracting energy from their environment, and they must learn how to do these things based on information stored and passed along in their genetic composition. According to what Crick called the *Dogma of Molecular Biology,* a cell functions and survives by making proteins working from information encoded in the chemical structure of its DNA. To accomplish this, the cell first reads (*transcribes*) the DNA into messenger ribonucleic acid (mRNA), which is then processed by a complex organic molecule (*a molecular machine*) called the ribosome—the cell's protein factory. Ribosomes take the amino acids they read in mRNA and string them together (in a process called *translation*) to form the proteins that are essential to life. Woese's reasoning was that every cell must have a ribosome in order to produce proteins and survive.

Right: A photograph of Carl Woese at the light table on which he analyzed X-ray films of ribosome structure using radioactive tagging and electrophoresis techniques. Each dark spot represents short pieces of specific 16S rRNA gene sequences.

Opposite: A hover of cut-throat trout spawning in a cold, clear and fast moving stream. These fish only survive in good clean water and the Yellowstone ecosystem is a model for preserving a healthy world where these trout can live and so can we. Growth rings in the otoliths or ear bones of the cutthroat record their life history and changing environmental conditions.

Therefore, to understand the evolutionary history of the cell, the chemical composition (*sequence*) of the various components of the ribosome must be targeted and comprehended. Woese summarized these findings and propositions in his 1967 book *The Genetic Code: The Molecular Basis for Genetic Expression*. He held that to derive biological function, especially of microorganisms, one must know the chemical structure, as well as the energetic motions and vibrations of the molecular chemistry of the ribosome.

As an incredibly complicated molecular machine, the ribosome contains nearly equal amounts of protein and RNA. Therefore, just as Darwin had sought to measure what he called *essential* evolutionary characteristics, Woese set out to measure the essential chemical compositions of the ribosomal RNA (rRNA) gene sequence. A fundamental idea about molecular evolution was that genes could be read like a clock, a concept first proposed by Crick in 1958 and later by Linus Pauling and Emil Zuckerkandl in 1965. The genetic code and its information molecules of nucleic acids will change at a consistent rate over an expanse of time, due to mutations, like the steady beat of a ticking clock. Studying the pattern of these changes in information molecules, whether they be amino acids or nucleic acids, is therefore like studying an evolutionary timepiece that never runs down. With this premise in mind, Woese believed that the rRNA gene was the ultimate chronometer with which to construct the universal tree of life. To learn to read this evolutionary clock, Woese labored in relative isolation, helped only by a small band of graduate students, postdoctoral researchers and colleagues. This kind of research is rare today, when large laboratory groups and collaborative networks dominate molecular studies.

In the early 1960s a new technique called *molecular sequencing*—laborious, complex, and even dangerous because of its use of radioactive materials—provided a leap forward for the Woese team and many others involved in DNA research. Originally developed at Cambridge by Professor Fred Sanger for protein analysis, this technique proved crucial for research into how DNA encodes for proteins; and in 1958 it earned Sanger the first of his two Nobel Prizes in Chemistry. When these cutting-edge analytical approaches were acquired and implemented at the Woese lab, they quickly resulted in a breakthrough. They soon found that the information-packed small subunit (SSU) of the rRNA was enough to detemine common ancestry and relatedness, although it would take months to collect a single SSU rRNA fingerprint. When sorted according to particle size settling time during centrifugation, the SSU rRNA is called the 16S rRNA gene sequence. The S unit ($10^{-13}$ seconds) is named after the Swedish chemist Theodor Svedberg, who was awarded a 1926 Nobel Prize in chemistry for developing the techniques of quantitative centrifugation of macromolecules. Sanger sequencing measured only short 13 to 15 nucleic acid base pair-long reads, while the entire 16S rRNA gene sequence is composed of more than 1,500 nucleotide base pairs. Woese collected and analyzed these short molecular sequences from as many different living organisms as possible after tagging them with radioactive phosphate ($^{32}P$) and allowing them to physically separate using two-dimensional gel electrophoresis.

Bison calves, like all infants, spend significant time playing. Playing develops strength, coordination and agility as well as bonding and competitiveness as a calf seeks a place in the herd's hierarchy. Calves play with each other, but rarely play with their mothers; this cow and calf played together for about five minutes. The cow stood with her head down, encouraging the calf to push against her. The calf backed up with her tail in the air and made short runs at her mom's head. The calf danced around thumping into the cow's large skull enjoying the game.

The world now knew of Woese and his universal tree of life. Strangely, however, this breakthrough resounded more strongly in scientific circles overseas than they did initially in the United States, and thus the first conference on Archaebacteria was held in Munich, Germany, in 1981. Later, in 1990, Woese and his colleagues Otto Kandler and Mark Wheelis formally proposed renaming the urkingdoms as the three domains of life: Eucarya, Bacteria and Archaea. This change drove an entirely new and unimagined wave of studies in taxonomy, phylogeny and evolution. As a result, the core assumptions supporting research and discovery in the biological sciences were permanently transformed.

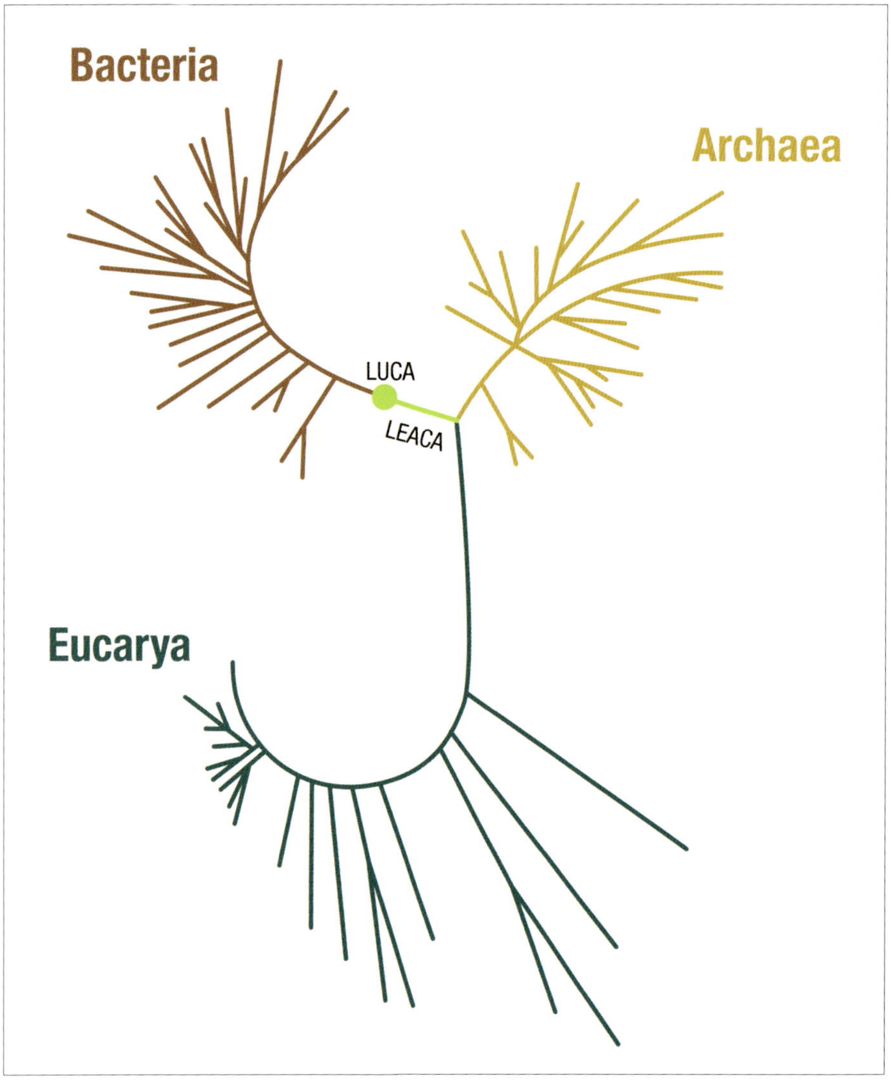


Right: An aerial view of the Woese universal tree of life *[modified after Pace, 1997]*.

Opposite: The Proximal-Slope Facies terracettes at Palette Springs show a diversity of microbial orange pigmentation and the encrustation and burial of pine trees.

Clepsydra Geyser, which means "water clock geyser," erupts nearly constantly in the Lower Geyser Basin. Dark clouds hang to the northwest ahead of a storm. The yellow deposits of elemental sulfur are the waste product of bacteria living in hot water.

## HOW DID LIFE EMERGE?

When Woese and Fox made these important changes to the tree of life, they provided an historical structure, an evolutionary connection among all life forms, past and present. This evolution unfolded over a vast expanse of geological time and shows that the Whittaker tree of life is but a recent twig at the tip of the Eucarya branch. To achieve an understanding of this enormous and complex drama, we have to follow more than one disciplinary path. We have to think about the geological history of the Earth, a history that extends back much farther than the story of the life on its surface.

How exactly did life begin? Solving that mystery has been an obsession for scientists and also a puzzle for imaginative writers, who engage with this question (or at least toy with it) every time they give us some variation of the scene where lightning zaps down a pole and a Frankenstein monster twitches on a slab. In the laboratories of real science, however, that wonderful or frightening moment still escapes us. When all of the DNA, RNA, proteins and other molecular components of a living cell are mixed together, we still have been unable to turn these cocktails into living cells. As far back as 1952,

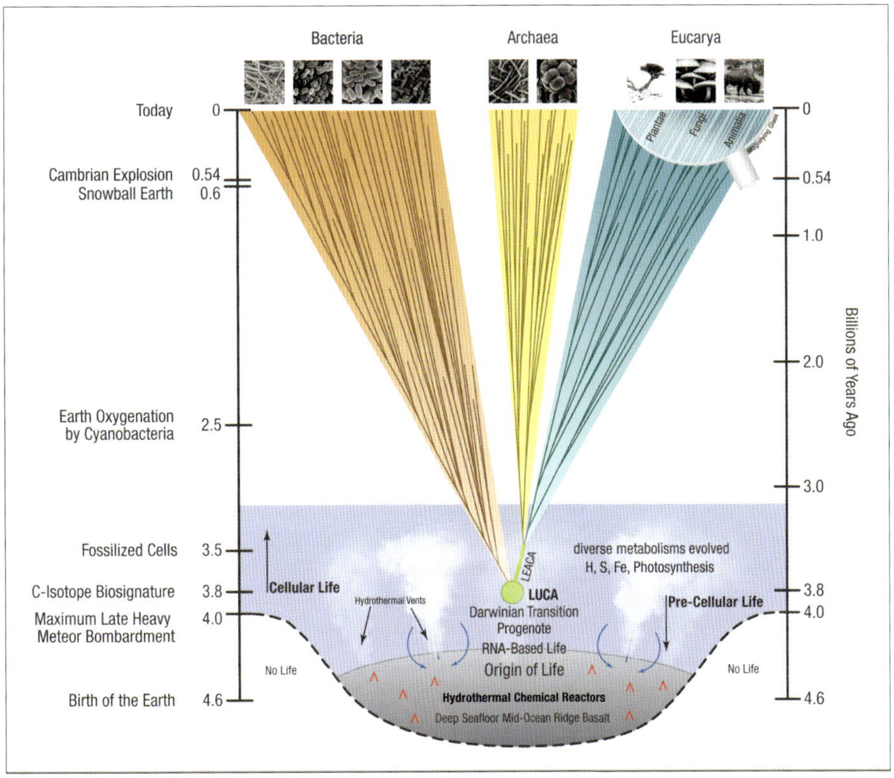


Opposite: Raven nest on a rhyolite cliff face in Golden Gate Canyon, named after the colorful lichen and rock. White streaks are guano from the nestlings. This nest site has been reused numerous times.

Left: The Woese universal tree of life in a framework of geological time and geobiological events.

Norris Geyser Basin green algae *Cyanidium* living in a low-oxygen and acidic hot-spring overflow channel. These algae are living at the uppermost temperature limit for photosynthetic life. The neighboring brown and red bacteria thrive in slightly cooler water.

In contrast to the shallow, warm, land-based ponds of "primordial soup" envisioned by Darwin and simulated in the lab work of Miller and Urey, it is now thought that life on Earth began in the more stable environments of the deep seafloor. Submarine mountain ranges called mid-ocean ridges rise kilometers above the deep abyssal plains of the world's major ocean basins. Hot molten rock rises from beneath the centerline of these mid-ocean ridges to cool and crystallize into the solid rock comprising the oceanic crustal plates, which are slowly moving apart and thus called ***seafloor spreading centers.*** One of the most extensive is the Mid-Atlantic Ridge, which runs north-to-south across the entirety of the planet from northern Greenland to the southern tip of South America. In this deep oceanic realm of the spreading centers, microbial life would have been sheltered from many of the direct effects of cosmic rays, solar radiation and even meteor strikes during the Late Heavy Bombardment. Immense volumes of cold seawater are constantly moving down and through fractures in the newly formed basaltic rock along the mid-ocean ridges. Such water-rock interactions create the highly acidic (pH 2 to 3) iron-sulfide black smoker vents we find today along the length of the spreading center ridges in oceans around the world. The water expelled from these vents reaches 350°C—a temperature too hot to sustain life. However, cooler off-ridge hydrothermal vents also occur along the length of these mid-ocean ridges. In these locations, the vent water is cooler and becomes more alkaline (70-90°C, pH 9-11) after circulating through and chemically reacting with the ultramafic igneous crust (low in silicon dioxide, high in manganese and iron), and with an overlying thin layer of marine sediments. These reactions produce huge volumes of diagenetically-altered rock deposits called ***serpentine***. In the process, seawater combines with the rich supply of iron available in the crust, releasing sulfur and the type of $H_2$-$CO_2$ redox couples thought to be essential for the origin of life.

At these common off-ridge vents, limestone chimneys have formed over the past 100,000 years, towering 60 m above the surrounding seafloor. These travertine deposits are, to a remarkable extent, a vertically-oriented version of the horizontal Apron and Channel Facies at Mammoth Hot Springs. They are built of aragonite and calcite crystals, as well as other minerals formed as a product of serpentinization, including double layer hydroxides (green rust), which rapidly encrust filamentous microbial mats. In the ocean, when the hot chemically-charged water flows up and out of the vent, it reacts immediately with the colder water all around, a process that drives the precipitation of iron and nickel. What results is an environment of rapid and extreme heat loss ***(thermal gradients)*** and chemical reactions that are quick and strongly unbalanced (far from ***equilibrium***). It is under these circumstances of turbulent carbon and energy exchange, where a return to physical and chemical equilibrium is so severely challenged, that life may have begun. From a chemistry point of view, the energy release and exchange in these contexts is like some whirling performance of Cirque du Soleil. However, instead of trapeze flyers, tightrope walkers and contortionists all in motion, we have a maelstrom of chemical reactions in which a variety of organic biomolecules can be spawned.

Right: This yellow-rumped warbler, whose body is the size of your thumb, is attracted to insects living in the thermal water at Mammoth. The bird's left foot is touching some spherules, the 1-2 mm sized travertine balls that form in the cooler pools where insects can survive and birds feed.

Opposite: A single aspen tree "suckers up" sending out rhizomes to create groves of genetically identical trees. Water and nutrients may be transferred between individual trees in response to the needs of the entire grove. Aspen groves form the largest interactive organism known in North America, covering up to 106 acres with 47,000 trees from one root system and may be as old as 700,000 years. Effectively an aspen grove is one individual organism and in a sense a community of clones.

These related environments—sub-seafloor serpentinization and alkaline vent travertine precipitation—are therefore believed to be the natural chemical reactor vessels in which the biomolecules of life were first synthesized. In the process, tiny crystalline mineral chambers with microscopic channels are formed in these serpentine and travertine mineral deposits, forcing these molecules from the vent water to flow and react and thus create free-floating molecules that strongly resemble RNA. In this earliest ***RNA-based life*** there was no "individuality," because there were no separate and distinguishable cellular entities. Woese, along with Crick and Leslie Orgel, generated this controversial idea that nucleic acids played a central role in the origin of life. RNA-based life was a probable precursor to cellular life, with nucleic acids serving as both chemical catalysts and information transmitters. This hypothesis suggests that one of the best, most accurate basic definitions of life may simply be "an ability to evolve." As a result, instead of so much emphasis upon "Survival of the Fittest," a process that requires individuality, "successful" life should be more accurately thought of as an open and completely communal sharing of genetic information and of strategies for survival. All of this results in the basic capacity to evolve. Woese's work was the first to provide preliminary evidence that there was life, and also evolution, long before there were either fully formed genes or cells enclosed with an outer membrane.

## INTO A MICROBIAL WORLD

Carl Woese was confident that his findings on the evolution of the ribosome would revolutionize the core assumptions upon which modern biology is built and practiced, but sharing these results with the world got off to a rocky start. Because Woese had labored in anonymity for so many years and wasn't a regular presenter at professional conferences, he was largely unknown to his peers. Also, comparing oligonucleotide catalogs back then was nowhere near as straightforward and transparent a process as comparing DNA and RNA today. As a result, there was plenty of misunderstanding and skepticism about the validity of Woese's approaches. To make things worse, because the *New York Times* had made several errors in reporting the 1977 front page story, scientists read the piece as a misguided attempt by Woese to garner unwarranted attention. Complications were also fomenting with regard to the relative position of Whittaker's five-kingdom configuration, on which so much of biological research had been anchored. Woese showed that the entirety of Whittaker's findings, while extremely important, apply only to the relatively small upper twig-like portions of the Eucarya branch of the three-pronged universal tree of life—a proposition that was slow to win support among some practicing biologists.

There is clearly a lot of water escaping from the geothermally heated ground in the Upper Geyser Basin. Hundreds of columns of steam climb into the still early morning air. Some of these columns arise from geysers, some from hot springs, and some from fumeroles. The Firehole River runs northward through the Upper Geyser Basin gathering the runoff from all of them, carrying the warm water to the Missouri River and eventually to the Gulf of Mexico.

The predicament facing Woese can be seen in a long and remarkable series of exchanges he had with Professor Ernst Mayr at Harvard University, one of the foremost classical Darwinian evolutionists of the 20th century. These communications, published in 1998 in the *Proceedings of the National Academy of Sciences*, centered on Mayr's belief that the great morphological diversity of the eukaryotic world would overwhelm any molecular and biochemical diversity to be found in the microbial world. In other words, how could a few differences in the chemical composition of one single gene possibly measure the immense morphological diversity of animals and plants? For Mayr, molecular phylogeny was a competing approach to taxonomy—Woese's work posed a direct challenge to the assumption that the world of microbial life is a world of primitive simplicity. Mayr saw phylogenetic taxonomy as being distinct and separate from the study of evolutionary processes, which was generally thought to be the domain of geneticists.

In line with mainstream Darwinian thought, Mayr affirmed that the study of evolution should stay with the origin of species, and not concern itself with constructing a universal tree of life. Darwin had argued that in order to achieve a deep evolutionary phylogeny, one had to focus exclusively on the essential characteristics of an organism. His world was one of plants and animals, and it was explored by concentrating on describing directly observable morphological traits, the tail of the lemur, the beak of the finch. The best way to do this, he held, was to look at embryonic characters, because those characters would evolve gradually and could in fact remain stable and recognizable over an arc of evolutionary time. When Darwin was doing this research, no such information was available for microbes. Microbes did not have embryos or any other obvious trait that was useful for Darwinian classification. The primary means for studying microbes, the microscope, was confounded by the simple shape and structure of microorganisms.

Opposite: When looking at a mountain like Colter Peak in the Thorofare, we get a sense of how small and insignificant we are. Everything in this photograph and all matter is derived from stardust that originated in the cores of trillions of extinct and recycled stars. Ourselves, the mountain, the stream, the trees and the grass have collected here from particles that have travelled millions of light years. The universe and our Earth are constantly reshaped by remarkable combinations of processes. From the 1 mm grass seeds in the foreground to the 3,256 m high Colter Peak, we are seeing a billion-fold change in size. From Colter Peak, to our nearest neighbor the Andromeda Galaxy, the change in size is a billion-billion-fold.

Above: We come to Yellowstone looking for purity, values, beauty and the source of wild nature. Yellowstone is one of the few places that we can easily access and still find these things. On a trail sign, this black-billed magpie appears as a willing ambassador for the start of our visit.

But Carl Woese had the wherewithal to see things differently, and he did so with persistence and courage. While Darwinian morphologists use direct human observation, molecular phylogenists could investigate genes and biochemistry that are only detectable with new and sensitive instruments. Though Mayr and other classical evolutionists wanted to classify the prokaryotes as a single "monophyletic" group, Woese held that the 16S rRNA gene sequence that he and his colleagues had discovered proves that Archaea domain is as phylogenetically distinct from the Bacteria domain as it is from the Eucarya domain. Furthermore, because so much previous research in the 1950s and 60s had focused on one supposedly representative microbe, *Escherichia coli*, which could easily be cultured and manipulated in the laboratory, a kind of research myopia had set in, a groundless assumption that knowing everything about *E. coli* was sufficient for understanding prokaryotes in general—including the Archaea. Woese's molecular phylogeny, combined with the detailed biochemical analyses of Archaea like anaerobic methanogens, proved that this assumption was wrong. Even so, the scientific community continued to resist, as evidenced when Woese and Gary Olsen at Illinois were turned down for support to sequence an Archaea, on the grounds that because a sequence of *E. coli* already existed, that was enough: if you had seen one microorganism, you had seen them all. In fact, the only federal funding agency that was willing to invest in Woese's avant-garde ideas was the National Aeronautics and Space Administration (NASA). Without NASA's support it would have been virtually impossible to achieve his scientific goal of establishing a universal tree of life.

Humans arrived in North America about 14,000 years before present. Elk have been an important food source since and continue to be. Cows and calves were walking up the flood plain of the Lamar River while five mature bulls walked on an ancient river terrace above them.

This is an illustration of the fossilization of terrecettes, which were like those on the cover photograph. Collapse reveals the underlying structure and proof that the flat lying Pond Facies travertine to the right is the base that was overgrown by pasta-like travertine on the left.

# DEEP TIME

*Time travels in diverse paces with diverse persons.*

—WILLIAM SHAKESPEARE, 1564-1616

## TIME KEEPERS

MARVELING AT THE RECURRENCE OF THE SEASONS, the sure motions in the heavens, the inexorable, onward rush of earthly days and years—human beings long ago adapted their cultural life to respect and conform to what they knew of the mysteries of time. Centuries before the first Egyptian pyramids, astronomers in Mesopotamia were contriving calendars based on the phases of the moon. In ancient Rome, a collegium of high priests ruled not only on matters of religious law, but also on the calendars of the Republic and the Empire, with authority to tamper with the weeks and the months when corrections were called for. It is clear that long before Isaac Newton first proposed in his 1687 *Philosophiae Naturalis Principia Mathematica* that time is a constant, people everywhere trusted in the accuracy of their sundials and stone circles, their lunar charts and mechanical timepieces.

Abiathar Peak at sunset in the Absaroka Range near the northeast corner of Yellowstone. The horizontal lines, visible here in this spot of golden light, illuminates multiple 34 to 56 million year old Eocene volcanic lava flows.

In the Western world, the phenomenon of the swinging pendulum was discovered by Galileo and adapted soon after as a fundamental mechanism of modern ticking clocks. This must have evoked in countless citizens a feeling of reassurance, and also of remorse, a steady unrelenting process carrying us forward and also bearing us all, as F. Scott Fitzgerald said, "ceaselessly into the past."

When long distance travel became much easier and more popular with the advent of railroads and steamships, the time of day in places that were once unimaginably far from one another, had to be brought into some kind of accord. In 1884, Greenwich Mean Time (GMT), named for the pre-eminent British observatory near London, was established as a kind of global king post for setting the hour in each place according to its longitude on the Earth. In the century that followed, extraordinarily precise time-measuring systems like the cesium atomic clock were developed, to measure time with an accuracy within one second every thirty million years. More recently, Coordinated Universal Time (UTC), as defined by an array of hundreds of atomic clocks, became the accepted standard for calibrating many operations in our high-tech life, from international stock market exchanges to international air traffic control and the Global Positioning System (GPS). If this dedication to nanosecond accuracy seems obsessive, it may also be natural, in a sense. The biological rhythms that keep us alive—heartbeat, metabolism, patterns of sleep and even of gene mutation—are themselves finely tuned clocks.

With all this measuring, and with centuries of dependence on Newton's propositions, most of us take it for granted that the passage of time is consistent everywhere on our planet, as well as throughout the universe. Yet this assumption turns out to be wrong. Consider the following. No matter how you observe light—no matter where you are or how fast you yourself are moving, in a hurtling spaceship or a whirling planet—its speed, absolute or relative to you, is always the same. Nothing else in the known universe operates by such a strange rule and this single insight, attributed to Albert Einstein, required a whole new understanding of how space and time are related to each other. Einstein's theories of "special and general relativity," first published together in the early 1900s, showed us that time does not flow evenly and smoothly forward throughout the universe. Instead, the flow of time can actually change depending on the position and velocity of a given observer or the dynamics of matter and energy. Hard as it might be to imagine, we have all heard a related truth unknown through most of human history: the faster you move, the slower time goes—each of us lives in our own time bubble. Space and time are no longer imagined as separate, but rather as two aspects of a single entity called space-time. We used to think of reality as three dimensional, but now it has four dimensions, with time as the fourth. Furthermore, we do not have to be on a starship to see the consequences. With a difference of about one billionth of a second, time flows more quickly on Earth than it does for the satellites currently in orbit running our GPS systems. Therefore, corrections for this time differential have to be made constantly, in order to assure that calculations of global positions will not go awry. Accordingly, as we venture to other planets, necessary calibrations like this will need to be highly sophisticated, to keep our control centers "up" with our astronauts, and also to keep those explorers alive.

The smooth soft delicate winter coat of a red fox stands in stark contrast to the heavily fractured and lichen coated volcanic rocks. The small, cold snow drift lies on rock that was once molten bedrock at 1,000's of degrees Celsius. This time and space continuum demonstrates that, given enough time, environmental conditions in the same place can change radically. These red fox were introduced to North America only about 200 years ago, while other fox species are native to North America such as the kit, swift and arctic fox.

## THE BEGINNING OF TIME

Theoretical mathematical calculations, combined with astronomical measurements of cosmic microwave background radiation (CMB, the most ancient light that reaches our telescopes and sensors), suggest that the universe was born in a "Big Bang" event approximately 13.8 billion years ago. According to NASA scientists and many academic researchers, the entire universe at that time was compressed into an unimaginably hot and dense accumulation of matter only one cubic centimeter in volume, infinitely denser than any black hole we have discovered and at a level of compression from which no light could escape. In the first moments of the Big Bang, the universe expanded tremendously, doubling in size in every nanosecond and reaching temperatures around 10 billion degrees Celsius. The background radiation we can pick up now is an afterglow from this initial cataclysm. This early inflation quickly created the universe basically in the form that we see today. For several billion years after, stars, galaxies and planets took form, expansion gradually slowed, an effect of gravitational attraction. The recent discovery of gravitational waves has been understood as strong evidence for the Big Bang hypothesis, but there is still so much that we do not know. For example, "dark energy," about which almost nothing is currently understood, apparently makes up more than half of our cosmos and is accelerating the expansion of the universe. It is assumed in our time that time itself is infinite, but that space somehow is finite and limited, and that "empty" space actually has a measurable temperature. Distant galaxies in the future may eventually be moving away from the Earth at the speed of light, which means that no starlight at all will be visible from our planet. It is also possible that a "multiverse," a possibly infinite array of different universes in different space-time configurations, was formed during the Big Bang, and that our own universe has literally no end of company.

Within minutes of the Big Bang, hydrogen (H) and helium (He) atoms came into being, and the universe expanded from 1 centimeter in diameter to 50 million kilometers across. Within five minutes, new atoms and molecules had accumulated into vast clouds, ***nebulae***. About 200 million years later, the mass of some of these nebulae had grown large enough for their gravity to pull in additional matter. As their density increased, these clouds morphed into swirling disks, rotating with rising velocity around a core, a spinning ball compressing and heating as gravity did its work. As the process continued for another 200 million years and these cores achieved temperatures of 10 million degrees Celsius, fusion reactions were triggered, and the first true stars were born. These stars in turn gave birth to the rest of the heavier atoms (carbon, sulfur, gold, etc.) that make up the Periodic Table of the Elements and launched them back out as ***solar wind***. The larger the star, the faster it burns. All stars eventually meet their demise as ***supernova*** explosions, expelling their accumulated gases and elements back into space.

So how does all this relate to Yellowstone and to Mammoth? All of the elements that make up the cells of all the life in the Park, and all across the planet, were created in those opening nanoseconds

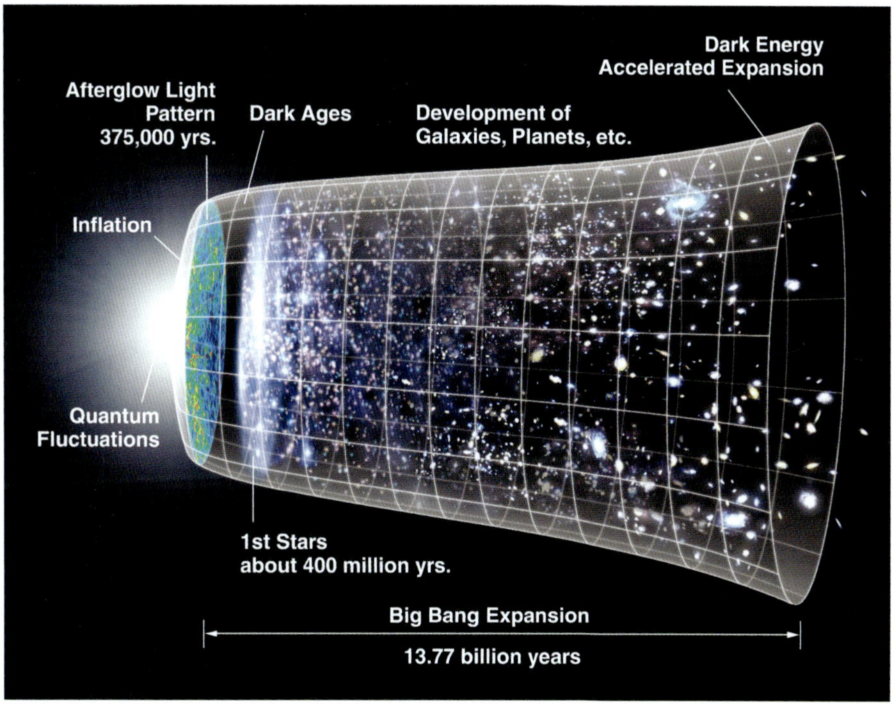


Timeline of the Universe, a graphical representation of the origin and evolution of the universe, developed by the NASA Wilkinson Microwave Anisotropy Probe program.

and during the billions of years of the history of the universe. Those rotating disks gave rise to galaxies and stars, collisions and agglomerations of dust, gas, rock, metal, and various kinds of ice (frozen water, carbon dioxide, and methane). This in turn created other celestial bodies such as asteroids, comets, and planets like our Earth. Our Solar System, located in an outward spiral of the Milky Way galaxy, is thought to be about 4.6 billion years old and 30 billion kilometers in diameter. These estimates are based on analysis of rocks and metals that have fallen to Earth as meteorites, and on extrapolations based on the oldest stars we can detect (their age, and also their velocity and rate of acceleration away from us), calculations that take us all the way back to the Big Bang. Planets originate when these accumulations of matter are heated by the energy of countless impact collisions, by gravity concentration and also by the decay of radioactive elements. When the size and mass of a planet becomes large enough, the interior can differentiate into a metal-rich *core*, a thick molten layer surrounding it called the *mantle*, and a thinner outer shell of solid rock called the *crust*. Planets become spherical because their gravitational force has grown strong enough to cause their rock, both molten and solid, to move, flow and equilibrate, an adjustment process that is nearly the same in all directions. Analysis of rocks collected from the Moon's surface by the NASA Apollo missions in the 1960s and 1970s, combined with the analysis of impact craters on its surface, suggest that an early planet-like body about the size of Mars slammed into the Earth at about 4.4 billion years ago (Bya). The debris from the blast, which was ripped from the crust and mantle of the Earth by the impact, then became an orbital ring that quickly coalesced to form the Moon.

All of this work goes into the determination of relative ages—where the timing of one geological event is compared to others—as well as into the kind of dating that we think of as absolute where we assign a specific time or age. Altogether, this work has established a firm and well-documented history of our planet, an essential framework for scientific discovery in Yellowstone and around the world. Though the story is rich and has many dimensions, it is a narrative that anyone with patience can follow and understand. We can even represent its major plot lines in pictures and charts. One problem with it, a trouble that puts off many people who might otherwise be fascinated, is the proliferation of strange-sounding names that have been assigned to various periods, lots of Greek and Latin roots, and adjectives that have been cooked up to incorporate some location where a discovery was made. For instance, the *Devonian Period* is named after rocks exposed in Devon, England, while the *Jurassic Period* is named for the European Jura Mountains. That tradition, and the enormous reaches of time that we need to remember, can seem intimidating, but we can get beyond those obstacles. For starters, here is a simplified concept map of geological time, showing a few of the sentinel events and ages in the history of the Earth. For geobiological "citizenship" on planet Earth—an understanding of where and when and what we really are—this can be a place to start.

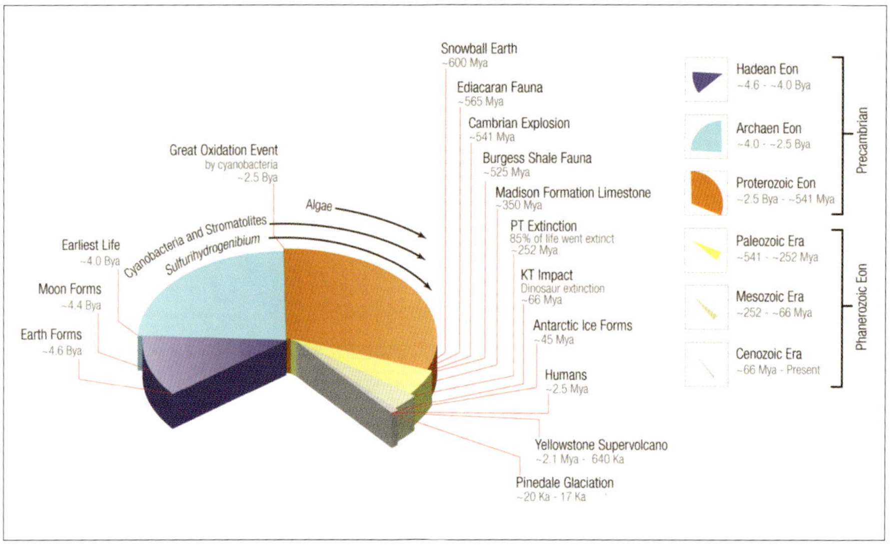


Overview of the geological history of Earth since it formed 4.6 Bya.

Almost ninety percent of our planet's history unfolded before the appearance of multicellular organisms, the Eucarya. This immense overture, a time interval known as the Precambrian Period, includes three eons: the Hadean, Archean and Proterozoic. In the Hadean Eon (appropriately named for Hades, the dark, forbidding Underworld imagined by the Greeks and Romans), free oxygen was virtually absent. This key component of life gradually increased its presence in the atmosphere as the Precambrian proceeded.

Delicate rice paper thin dragonfly wings have already accumulated travertine in less than 24 hours. Notice the brown microbial pigmentation of the pond travertine shrubs are also on the outside perimeter of the insect.

Oxygenation of the Earth was fostered by microbial photosynthesis, a process that evolved in cyanobacteria and was later assimilated by algae, and eventually by plants. The most recent half-billion years of Earth history is called the Phanerozoic Eon, derived from ancient Greek meaning ***visible life***, and composed of the Paleozoic, Mesozoic and Cenozoic Eras. These divisions of the Phanerozoic, defined in large part by crucial biological and geological events recorded in the fossil record, is one of our best guides to animal and plant originations and extinctions. There were no human beings until about 2.5 Mya—thus we have only just arrived. Bacteria like those in the hot springs at Mammoth, have been thriving a thousand times longer than we have. To place these events into more practical chronological terms, consider the entire 4.6 billion years of Earth history as equivalent to one 365-day calendar year. In this context, the Precambrian spans from January 1 through late October, dinosaurs reigned for a couple of weeks from late November through early December, humans did not appear until the last 5 hours of the last day of the year, and the United States Declaration of Independence was not signed until the last second of New Year's Eve.

One school of geological thinking has it that until about 4.3 Bya, the planet was a hostile and violent place—a seething ocean of molten rock *(magma)* with a thin skin of solidified rock forming, sinking, and re-melting. Add to this turmoil a barrage of meteors, some of them up to 150 kilometers in diameter, which frequently hit the surface with devastating effects. All of this turmoil and change of physical state would have impeded the supply of stable liquid water, an essential ingredient for life. Other scientists, however, see some exaggeration in this vision of the planet as a constant seething ocean of magma, and propose instead that even a planet experiencing many volcanic eruptions and meteor impacts is still likely to have cooler, more solid crust most of the time. This dissenting view is supported by oxygen-isotope analyses of 4.4 billion year-old zircon crystals from sandstones in the Jack Hills region of Western Australia, suggesting that the Earth may have been cooler and wetter even at this early stage. Either way, gasses released from these molten expanses and outpourings accumulated to create a toxic atmosphere. As water in this atmosphere condensed and fell for long ages as rain, oceans came into being that were devoid of oxygen. The chemical compositions of the oceans and the air in the Hadean Eon, and in particular their ***anoxia*** (the oxygen absence) would have created strange panoramas, very different in color from what we see today.

As the Hadean Eon segued into the Archean (4.8 to 3.8 Bya), life is thought to have emerged in ridge spreading centers at the bottom of the seas. At about the same time, meteor impacts on the Earth's surface reached their height, an intensity referred to as the Late Heavy Meteor Bombardment, evidenced by patterns of craters we can observe on planets and moons throughout the solar system. Soon after the advent of the Last Universal Common Ancestor (LUCA), several ancestral forms of bacteria originated, many of which have living descendants today. This includes the Aquificales lineage of ***Sulfurihydrogenibium yellowstonense***, the filamentous sulfide-oxidizing bacterium that dominates the Apron and Channel Facies at Mammoth. Banded iron formations (BIFs), a hallmark of the Archean seafloor and now actively mined for steel production, are composed of thin recurring layers of dark iron ore ***(hematite*** and ***magnetite)*** and various types of red chalcedony ($SiO_2$). Bacteria in these deep marine settings evolved the process of ***anoxygenic photosynthesis***, which used electron sources such as dissolved iron ($Fe^{2+}$). Though this kind of photosynthesis did not produce oxygen, it may have played an important part in the creation of these BIF iron ore formations. A complex biochemical transition then took place in the Archean when cyanobacteria such as *Synechococcus* and *Chloroflexus*, which now inhabit the Pond Facies at Mammoth, evolved the capability of ***oxygenic photosynthesis***. This metabolic process combines $CO_2$, water ($H_2O$) and sunlight energy in the presence of oxygen to produce sugars ***(carbohydrates)***. Pigments such as chlorophyll were utilized to split the water, produce oxygen ($O_2$) and thus transform light energy into chemical energy. In other words, photosynthesis allowed electrons from the water to be "donated" toward the reduction of $CO_2$ and the ensuing formation of food. Once that oxygen began to be supplied via cyanobacteria photosynthesis, it would have quickly reacted with dissolved iron in seawater to further enhance BIF deposition.

Magpie tracks preserved in a layer of hollow balls of travertine that are encrusted gas bubbles less than a millimeter in diameter. They form when $CO_2$ gas bubbles float at the surface of flowing water and become coated with travertine within seconds. When fossilized, it is called foam rock.

By about 2.5 Bya, the Earth's oceans and atmosphere were charged with oxygen derived from photosynthesis, creating the worldwide Great Oxygenation Event (GOE) that marked the beginning of the Proterozoic Eon. From the Greek words *proteros*, meaning "first," and *zoon*, meaning "life," the Proterozoic was ironically actually the second age of life and is characterized by cyanobacterial and algal Eucarya *oxygenic photosynthesis*. This ongoing introduction and increase of global oxygen profoundly influenced the evolution of life on this planet, helping to drive the radiation and diversification of all three domains of life. One of the earliest thermophilic algae to evolve in the Proterozoic was the unicellular red algae *Cyanidium*, which still thrives in highly acidic, $CO_2$- and $SiO_2$-rich hot springs in the Norris Geyser Basin of Yellowstone. However, it is absent from the cooler, neutral pH and $CaCO_3$ precipitating hot springs of Mammoth.

Finely layered *(laminated)* mounds of sedimentary rocks, called *stromatolites*, began forming in the Archean and later proliferated in the Proterozoic. Some of these stromatolite deposits reached a kilometer in thickness as they accumulated on sinking seafloors. These provide a complete record of this primordial world of microorganisms. Well-preserved ancient examples of these thick formations are found in the region of Western Canada that geologists refer to as the Slave Province. This process continues even today in Shark Bay of western Australia, the Bahamas and the Maldives. Tiny particles of sediment adhere to sticky, mucus-rich cyanobacteria mats and slowly yet actively accumulate layer after layer. Evidence from fossils and isotopes tells us that this layering of life and mineral was a dynamic of ancient Archean and Proterozoic seafloors. However, similar deposits can also sometimes take shape in the absence of microbial life, thus not all stromatolites are biological in origin. BIFs started in the Archean and continued to be deposited on the Proterozoic seafloor. Around 600 Mya, ice covered all of the world's oceans, including the equatorial zones. This "Snowball Earth" event may have caused seawater oxygen levels to drop so low as to impact bacterial life. Ironically, a rise in what we now call "greenhouse gases" turned out to be crucial in ending the dominion of snow and ice and bringing the planet back to warmth and vibrant life. At the very end of the Proterozoic era, lineage of multicellular Eucarya called the Ediacaran fauna, named after the Ediacara Hills, were discovered near Adelaide, Australia, and had no skeletons, flourished briefly and soon went extinct.

Very recently in the history of the Earth, only 542 Mya, something utterly new took place. Called the *Cambrian Explosion*, an enormous variety of multicellular Eucarya suddenly evolved on the seafloor of oceans around the world. Though the exact reasons for the Cambrian Explosion remain uncertain, it may owe much to the high levels of oxygenation that the Earth had attained by this time, as well as to ecological advantages achieved with larger size and skeletal armor. When you get bigger and tougher, you get better at fending off predators. This first occurrence of such large multicellular organisms *(metazoans)* with external skeletons *(invertebrates)* marks the beginning of the Phanerozoic Eon. Age subdivisions within the Phanerozoic are based in large part on the first and last occurrence *(radiations* and *extinctions)* of organisms deduced from fossilized skeletons and sediment imprints *(trace fossils)*.

The Mammoth Hot Springs area is a familiar home to elk, some of which spend time on the lawn in front of the Albright Visitor Center. In the active higher temperature Vent, Apron and Channel, and Pond Facies, it is dangerous for these animals because they may fall through into underlying pools and cavities. Generally, it is safest for these animals to occupy the areas away from any flowing water. Hot-spring bacteria clinging to the hair of elk, bison, deer, coyotes and other animals is a primary means for microbes to be transported throughout the Yellowstone ecosystem.

Summer evening in the Lamar Valley looking east to the Absaroka Range. Specimen Ridge, on the right side of the photograph and south side of the valley, contains 29 successive layers of petrified forests from the Eocene. Diverse genera of silicified magnolias, avocadoes, sequoias, palm trees and other plants now native to the southern U.S., have been well preserved. This Eocene volcanic bedrock has been successively reshaped by glacial and river events and processes, resulting in a beautiful high mountain landscape.

## LAND OF FIRE AND ICE

The chronological history of Yellowstone hinges upon two powerful forces in nature—the explosive eruption of volcanoes and the unrelenting march of glacial ice. About 16 Mya, the North American Plate of the Earth's crust began moving southwestward over a stationary deep Earth mantle "hot spot" plume. This slow, immense displacement of crustal rock above an unimaginably huge column of upward moving molten rock, like a giant lava lamp, has left behind a four hundred kilometer-long arc-shaped wake of volcanoes and lava flows creating what is called the Snake River Plain. The Snake River Plain is a series of volcanic calderas, which start at the Nevada-Oregon border, becomes progressively younger toward the northeast, and eventually ends up at modern day Yellowstone. When the hot spot is directly underneath, the crust is heated and melted, topographic elevation dramatically increases, and vigorous volcanic activity takes place. For example, Yellowstone has been uplifted by as much as 1 km during the past five million years, and now stands high in elevation above the Snake River Plain. As the crust moves beyond the hot spot, volcanic activity eventually ceases, the crust cools and becomes dense, and the land surface subsides to lower elevations. Millions of years in the future, as the North American Plate continues to drift, the dynamic that created Yellowstone will shift to eastern Montana and the Dakotas. A similar process is underway in the Hawaiian Island Chain except under oceanic conditions, created and constantly transformed by migration of the Pacific Plate over a stationary mantle hot spot.

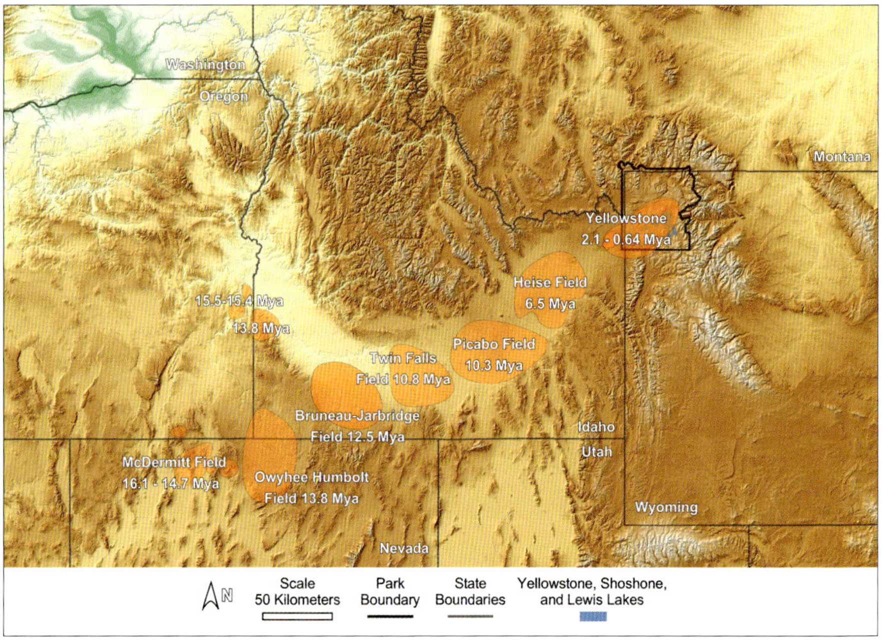


Regional map of the Snake River Plain volcanic track. The curved shape was formed by the southwest-ward migration of the North American Plate over the Yellowstone hot spot. The original eruption was about 16 Mya in the old McDermitt volcanic field on the Nevada-Oregon border, with successive eruptions along the Snake River Plain that end at the Yellowstone volcanic plateau.

The 13,000 year old Pinedale glaciation bulldozed rocks and soil toward the northeast to help shape the Blacktail Plateau just 10 km east of Mammoth. This dramatic rainstorm is contributing to the hydrothermal system of Yellowstone.

By measuring the travel time of energy generated from earthquakes at points around the world *(seismic tomography)*, scientists have ascertained the geological structure deep beneath Yellowstone. Continental crust is composed of $SiO_2$-rich igneous rock, which melts and mixes with the metal-rich hot spot magma that is moving up from below. This motion creates a shallow horizontal magma reservoir within the crust that is rich in silicon dioxide, less dense and much more explosive than the deeper chambers being fed by magma. To reconstruct the volcanic history of Yellowstone, volcanic rocks were mapped within and around the Park in the 1960s by Bob Christensen and his colleagues at the United States Geological Survey (USGS). ***Basalt*** is a dark metal-rich igneous volcanic rock with very small crystals that forms when deep mantle-derived magma comes directly to the surface in lava flows as we see underway on the island of Hawaii today. Rhyolite, a silica-rich igneous rock, is made up of relatively small mineral crystals. These were deposited from explosive eruptions of the shallow magma chambers directly beneath the caldera. ***Rhyolite*** makes up the core of the Yellowstone Plateau that is covered by vast expanses of forests comprised almost solely of lodgepole pine *(Pinus contorta)*. ***Tuff***, also called an ***ignimbrite*** or ***pyroclastic flow***, is a deposit of fragmented volcanic debris. Tuff is deposited as red-hot debris races down the volcano's side like an avalanche, as well as from volcanic ash settling out of the air like snow. Tuff forms widespread thick layers that weld together from the heat, condense, cool and become heavy over time. Pompeii was incinerated and entombed by similarly lethal pyroclastic flow deposits that were violently erupted from Vesuvius.

Opposite: Looking northwest from Swan Lake Flat just after a September sunrise at Electric Peak in the Gallatin Range near the northwest corner of YNP. Gardners Hole lies here between Electric Peak and the ridge with scattered grass and timber in the center of the photograph. Rain and snow runoff primarily flows eastward off the backbone of the Gallatin Range into the fault zone that creates Gardners Hole. The water then migrates down into the warped Madison Limestone Formation, is heated by the Yellowstone supervolcano, and dissolves the bedrock. The chemically saturated water returns to the surface to form the Mammoth Hot Springs travertine formation.

Yellowstone has had three separate major volcanic eruptions over the last two million years, each of which formed a central volcanic crater (caldera) and was accompanied by hundreds of smaller intermediate eruptive events. The first and largest eruption occurred 2.1 Mya, expelled 2,500 cubic kilometers of ash and rock debris, and is recorded by the Huckleberry Ridge Tuff whose associated rhyolite forms the cliff faces at the Golden Gate just a few kilometers southwest of Mammoth. The second and smallest eruption happened at 1.3 Mya as represented by the Mesa Falls Tuff. This generated 280 cubic kilometers of ash and can be seen filling eroded lows within the Huckleberry Ridge Tuff cliffs exposed at the top of Mt. Everts to the east of Mammoth. The third, yet second largest, major eruption occurred at 640,000 years ago, during which the Lava Creek Tuff was deposited. More than 1000 cubic kilometers of ejected material was removed from underground, which would be enough to bury the entire state of Alaska in several meters of ash and rock debris. When this much material is removed during an eruption, the upper surface caves in and forms a caldera with many fractures in the surrounding ground surface. The elliptical Lava Creek caldera is approximately 80 by 50 kilometers in area and is the site of thousands of earthquakes annually. Two large areas within the caldera, called resurgent domes, are preferentially uplifting at variable rates.

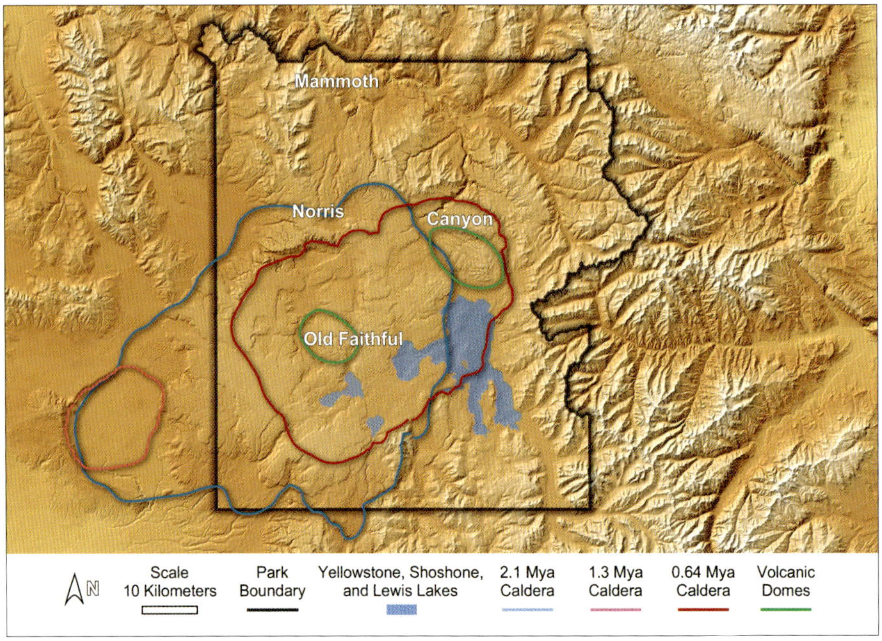


Right: Regional map showing the geographic position of the three Yellowstone supervolcano calderas. The blue outline is the 2.1 Mya Huckleberry Ridge caldera, the pink outline is the 1.3 Mya Mesa Falls caldera, and the red outline is the 640,000 year old Lava Creek caldera. The green outlines within the Lava Creek caldera are the uplifting Mallard Lake Dome and Sour Creek Dome.

Yellowstone is therefore the world's largest violently erupting continental "supervolcano," having cumulatively expelled some 3,780 cubic kilometers of ash, rock debris and lava. Utilizing the Powers of 10, this is more than the material ejected from the Mount Saint Helens, Mount Pinatubo, Krakatoa,

Mount Mazama and Tambora volcanoes, combined. Mauna Loa, a Hawaiian seafloor volcano, has expelled the largest total volume of erupted basalt from a single volcano at about 80,000 cubic kilometers. Even larger in total volume are the enormous lava flows such as the continental Columbia River Basalt and oceanic Ontong Java Plateau (large igneous provinces), which are cumulatively thought to represent a total volume of as much as four million cubic kilometers of extruded mantle-derived basalts.

So what can we expect from the Yellowstone supervolcano in the near future? The straight answer is that no one knows for sure. However, decades of scientific research coordinated by the Yellowstone Volcano Observatory leads to the following speculations. The next major eruption at Yellowstone will probably not be a catastrophic event that destroys all of North America. Instead, it is likely to be either a hydro-thermal eruption of rock and hot water, or it will be a non-explosive basaltic lava flow. In either case, the next eruption event will likely have relatively minimal effect outside of the Yellowstone region. Furthermore, the magma chamber below Yellowstone is not growing larger and it is not moving. This is because most of the magma chamber is only partially melted rock that is not expanding. Therefore, not enough melt is available to fuel an explosive super eruption. Although still seismically active, recent earthquakes are associated with the type of low-frequency energy that is linked to deep crustal magma, which is the kind that produces lava flows. It also does not appear that the next big eruption is overdue. The average interval between eruptions at Yellowstone is well over 700,000 years, and the last major explosive eruption was 640,000 years ago. Even so, this is based on an extrapolation from a limited set of eruption events. It is worth noting that the last significant lava flow eruption in Yellowstone happened 70,000 years ago. In addition, the surface of the resurgent domes within the caldera, previously recorded to move upward 70 cm, have not risen further since 2015, suggesting that activity may be diminishing. If only there were a geological crystal ball with which to predict the future of Yellowstone eruptions.

Some of the most breathtaking mountains and valleys in Yellowstone were sculpted by glacial ice. Deep snowfall over the last 200,000 years (the ***Pleistocene Epoch***) resulted in more snow falling in the winter than melted in the summer. This was caused by the high elevation of the Yellowstone Plateau, climatic effects induced by the Snake River Plain, and cooler global temperatures. These factors resulted in two primary episodes of glaciation that covered all of Yellowstone and formed the largest mountain glacier system in the western United States. The first portion of each interval saw glaciers initially form and flow down valleys from the high mountain plateaus. As glacial ice accumulated, it in itself increased elevation and had a self-perpetuating effect that further enhanced ice accumulation. In the second stage of each glacial episode, these glaciers reached their maximum size to form one large icecap covering nearly all of Yellowstone and the Grand Tetons.

This is what an active glacier looks like on South Georgia Island, Antarctica. These modern day glacial processes, which create distinct landforms like these jagged arêtes, horns, and cirques, are used to interpret the glacial geomorphology from prehistoric times.

Bighorn sheep, like these full curl rams, still survive in Yellowstone in relatively small numbers (< 300 in the Park). Prehistorically they were much more common. There are proto-historic accounts of a Shoshone language group called the Sheep Eaters living year-round on the Yellowstone Plateau. The male on the right-hand side is wearing a radio collar, part of one of many ongoing scientific studies in Yellowstone.

The timing of Yellowstone glaciations has been pieced together using multiple lines of evidence collected from throughout the Yellowstone region, including the mapping and stratigraphy of glacial deposits, rock chemical weathering analyses and absolute isotopic age dating. This work suggests that the older Bull Lake glaciation, which covered all of Jackson Hole and western Yellowstone, reached its maximum at about 136,000 to 150,000 years ago. The later Pinedale glaciation was most extensively developed at 17,000 to 20,000 years before present, reaching a thickness of more than 1 kilometer and an upper surface elevation of more than 330 kilometers above sea level. In northern Yellowstone near Mammoth and Gardiner, multiple thick glaciers converged to form an outlet glacier that carved the Paradise Valley. The head of this glacier was poised near the mouth of the Gardner River and flowed to the north where thick lateral and terminal moraines (sediments at the margins of the glaciers) were deposited. The land surface shapes and sediments resulting from the Bull Lake and Pinedale glaciations have a profound influence on the Yellowstone ecosystem to this day.

Because nights in Yellowstone are largely free of the ambient human light that clutters the sky in so many other places, the Park is a wonderful place to look at the heavens. One nearby mountain peak is named for a scientist who did much to help us understand what we are looking at, both on Earth and overhead. In 1859, a German chemist named Robert Bunsen, working with a physicist named Gustav Kirchhoff, invented spectral analysis, a technique that allows us to determine the chemical and physical properties of distant celestial phenomena, including their temperature and velocity through space. Bunsen Peak, immediately south of Mammoth, was named for him to honor his work on geyser activity, but as we look around us, it is worth remembering that his curiosity and achievements extended to the entire universe. When you are sitting quietly on a hillside in the Lamar Valley, remember that the Earth is spinning to the east at almost 1,450 kilometers per hour and at the same time is flying at 150 kilometers per second in its orbit around the sun. Our own small star and its scattering of planets are all moving at about 1.6 million kilometers a day in a grand orbit around a black hole at the center of our galaxy, making the circuit once every 200 million years. We are going someplace in the dark cold abyss of space, but we do not know exactly where, and it does not really matter. If you feel a little dizzy thinking about all this, it is normal. Yet it is worth remembering that we must concentrate our attention right where we are now, on taking care of our home, this small and unique speck of glorious life and beauty. We need to treat all of our neighbors—from bacteria to blue whales—with kindness, consideration, generosity, and humility.

The carefree playfulness of this mother grizzly bear and her two cubs illustrate the security and value of preserving a wild place like Yellowstone. This equally applies to preserving the wildness of Yellowstone geology, geochemistry and microbiology in places like Mammoth Hot Springs. There used to be another species of bear in Yellowstone that was 50% larger than these grizzlies, called the short-faced bear, which went extinct about the time that humans arrived at the end of the Pinedale Glaciation. We cannot afford to see places like Mammoth Hot Springs go extinct like the short-faced bear and other creatures that once thrived here.

A gold sunrise during an eruption of Old Faithful on a cold winter morning. The ball of the sun is gleaming through a 150 meter-high column of microbe and mineral laden steam. The gold color is from particles in the air, and the water is steaming at the bottom from the 3,000 gallons of runoff released during each eruption. The roofline is of the Old Faithful lodge that sits just east of the geyser cone. The geothermal hydrology that causes geysers can only happen in young volcanics like these silica-rich rhyolites, which can handle the high pressure. Cone type geyers like Old Faithful erupt through a nozzle in the rock, while a fountain-type geyser erupts through a pool of water. There are more than 300 geysers in Yellowstone, which is more than half of the known geysers in the world

Detail of the Lewis and Clark Map shows the route of Colter's walk in 1807-1808, indicated by a dotted line on Clark's map. After he returned to St. Louis in 1810, Colter personally gave Clark this information. Colter's walk was the first known European visit to what later became Yellowstone.

# HUMAN HISTORY

## EARLY INHABITANTS

FOR THOUSANDS OF YEARS, human beings have played a significant role in the Yellowstone story. Archaeological work in Yellowstone has established that virtually all of the Park was occupied or utilized by prehistoric peoples for: quarries, vision-quest sites, butchering sites, habitation sites, and wickiup encampments, ranging in age from about 9,000 years ago to the first encounter with Europeans. The earliest sites were probably situated very near the edge of the retreating glacial ice, for there were advantages to occupying and using these "new" places as soon as they opened up. The wildlife came, and the humans followed. Another attraction was the availability of alpine and sub-alpine plants for food and medicinal uses.

Rapid
Ordway R.
Mountain
Judith
Muscle
Shell
Pryor's Cr.
Oat-la-shoot
430 Souls
Clark's
Shields R.
Snow
Otter R.
Dry Cr.
Fort Cr.
Gass Cr.
Howard's Cr.
Field R.
Wisdom R.
R. Across
Beaver R.
Brattan R.
Rosebud R.
Fork
Cap. Clark's
Jefferson's R.
Frasier's R.
Philosophy R.
Ap-sah-roo-ka R.
Clarks
Hot Spring
Sho-sho-nes
Souls
Philanthropy R.
Madison's R.
Gallatin's R.
Ne-moy
600 Souls
Band of Snake Indians.
Po-hah 1000 Souls
Band of Snake Ind.
Pass
Colter's route in 1807.
Yep-pe
Band of Snake Ind.
1000 Souls.
Lake
Southern
Wiser's R.
Henry's R.
Stinking Water R.
Mick-ka-appa R.
Boiling Spring
Heart Mn
Salt Fork
Big
Horn
Lake Biddle
Cas-ta-ha-na Tribe
1500 Souls.
Bands
RIO DEL NORT
Colter's R.
We-ta-pa-ha-to
1800 S
Rio de la Plata
Highest Peak
Block House U. S. Factory in 1806.

The Lewis and Clark Map, was the first relatively accurate map of the northern Rocky Mountains and the northwest plains drawn by William Clark. The map was published in 1814 in Nicholas Biddle's edited journals of the Lewis and Clark Expedition. This map served as the most reliable guide to this region for decades. Lake Eustis on this map, named for the U.S. Secretary of War, at that time, is now called Yellowstone Lake.

LONGITUDE WEST FROM WASHINGTON CITY
Northern Boundary of the United States Lat. 49° 37' N.
LAKE of the WOODS
Algonquins 600 Souls
Algonquins
Grand Portage
LAKE SUPERIOR
Red Lake R.
Red L.
Goose R.
Old Mandan Village
Teton
Saone
Ricaras 3000 Souls
Sisatoone a Tribe of Sioux 900 Souls
Big Stone L.
St. Peters R.
Red Wood R.
Prairie la Crosse
MISSISSIPPI
Rock R.
Foxs 1800 Souls
White R.
Quicurre R.
Mahar Vill.
Pawnee Loup Vill. 600 Souls
RIVER LA PLATTE
Great Pawnee & Republican Vill. 2000 Souls
Republican Fork
Solomons Fork
Kansas R.
Kanzas Vill. 1000 Souls
Old Kanzas Vill.
KANZAS R.
RIVER
ILLINOIS R.
LONGITUDE WEST FROM LONDON
G4126.S12 1814.L4 Vault copy 1
Division of Maps
Library of Congress
Force Map Coll no. 409 (Copy 1)

## EURO-AMERICAN CONTACT AND WESTWARD EXPANSION

The first documented journey into the Yellowstone Plateau by a Euro-American was made by John Colter, who stayed there in the autumn, winter, and spring of 1807-08. Colter had been one of the hunters enlisted on the Lewis and Clark Expedition, which had left St. Louis in May 1804, moving in the first year up the Missouri River to a place near where Mandan, North Dakota is now. Wintering there in 1804-05, they then successfully traveled all the way to the Pacific Ocean by November of that year, spending their second winter near the mouth of the Columbia River. On their return trip to St. Louis in 1806, floating down the Missouri River in present-day western North Dakota, they encountered two white men named Joseph Dickson and Forrest Hancock, who were fur trappers on their way up the Missouri River. In the few days they spent with the Lewis and Clark party, they convinced John Colter to break from the group, turn around, and accompany them back westward. Though Lewis and Clark were reluctant to see that happen—the Expedition was already a year behind schedule with many people back East assuming the entire company was dead—the leaders wanted to return quickly with as many skilled men as possible. Because the Expedition's members had enlisted for only a two-year stint, they were now under no obligation to stay on. Lewis and Clark gave Colter plenty of gunpowder and lead for his rifle, and some traps, and on August 17, 1806, they parted company, while the rest of the party continued all the way back to St. Louis.

*Golden Gate,*
*Yellowstone National Park,*
*Thomas Moran,*
*oil on canvas, 1893*

Thomas Moran's oil and water color art helped convince Congress in 1872 to pass legislation to create Yellowstone National Park. Moran was part of the Ferdinand Hayden expedition of 1871. His most famous paintings of the Yellowstone area were of the Lower Falls and the Grand Canyon of the Yellowstone. In 1893 he made this dramatic oil painting of Golden Gate, which is 8 km south of Mammoth Hot Springs. Without the visual power of art from Moran and photographer William Henry Jackson, and the existence of rare geothermal features, it would have been more difficult to persuade people to set this land aside from commercial development.

Nothing certain is known about Colter's adventures for the next year. At some point he went off on his own, for in the spring of 1807 a fur trader, Manuel Lisa, met Colter floating solo down the Missouri in a canoe loaded with pelts heading back towards St. Louis. Knowing that he had just found the best guide to the Upper Missouri, Lisa somehow convinced Colter to turn around once more and go back upstream with him. After Lisa erected a small trading post at the confluence of the Yellowstone and Bighorn Rivers, Colter went on a mission to inform the regional natives about this new place to sell their furs. On foot and carrying a pack and a rifle in the autumn of 1807, Colter headed up the Bighorn River, and returned in the late spring of 1808. In those intervening months, John Colter became the first white man to venture into what is now Yellowstone National Park.

The key record of Colter's hike across the Yellowstone Plateau, named Colter's Walk, comes from William Clark's detailed map in the ***History of the Expedition under the Command of Captains Lewis and Clark***, in a two-volume edition edited by Nicolas Biddle. On this map, though the circuit called "Colter's Route 1807" is clearly marked out, it is impossible to decipher which direction around the Park Colter actually traveled. His route was probably counterclockwise, allowing him to cross the Yellowstone Plateau in the autumn, assuming that he traversed Sunlight Basin to the head of the Clark's Fork of the Yellowstone, then journeyed down Soda Butte Creek and the Lamar River from Cooke Pass to the Bannock Trail crossing of the Yellowstone River at the mouth of Tower Creek—

The 1837 Stiles Map, published in Washington Irving's book *The Rocky Mountains: or, Scenes, Incidents, and Adventures in the Far West,* though inaccurate, was the standard for regional geography at the time. Some of Stiles' information for the map may have come from John Colter. In another one of Irving's books, entitled *Astoria*, he recounted Colter's run, which took place in July 1808. The first published account of Colter's run was in the Louisiana Gazette in 1809. Colter was trapping with Benjamin Potts, a friend from the Lewis and Clark Expedition. Colter and Potts were on the Madison River near the Forks of the Missouri when they encountered about 100 armed Blackfoot Indians. Potts tried to escape and was killed. They took Colter's weapons, stripped him naked and told him to run for his life. After a quarter of a mile or so, he had outdistanced all but one. Due to the stress of exertion his nose was bleeding and in desperation he stopped and turned to face the single man a few yards behind him. Covered in blood, he startled his pursuer causing him to stumble, and giving Colter the chance to grab the Indian's knife and kill him. Colter took off running from the remaining group and made it to the Jefferson River where he dove under a pile of driftwood at the head of a river island. The Indians could not find him, and after they left he walked barefoot and naked 300 miles east to the nearest fur post at the mouth of the Powder River where Miles City, Montana, stands now.

only 30 kilometers east of Mammoth Hot Springs. Subsequently, Colter went around Mount Washburn. He traveled up the Yellowstone River through the Hayden Valley and around the West Thumb of Yellowstone Lake, which appears on Clark's 1814 map as Lake Eustis. Colter probably trekked from there over the Continental Divide to the headwaters of the Snake River, into Jackson Hole and around the west side of "Lake Riddle" (now called Jackson Lake), and then into Pierre's Hole on the west side of the Tetons in present day Idaho, where he may have spent some of the winter.

On his probable return route, Colter passed over the Togwotee Pass west of Dubois, Wyoming, then along the east side of the Absaroka Range, and down the South Fork of the "Stinking Water

River," now known as the South Fork of the Shoshone. Yellowstone has been called "Colter's Hell" but almost certainly the array of sulphurous hot springs at present day Cody, Wyoming that Colter saw is the origin of that name, not the features in the present day Park. He did not travel farther in what became Yellowstone Park before returning to Lisa's fur trading post. Along the way he must have had local native help to navigate these difficult, complicated mountain ranges during five or six months of cold weather exploration.

The Rocky Mountain fur trade era in the Yellowstone region, which began in earnest with the opening of Lisa's trading post, lasted from 1807 to about 1840. By 1840 the population of beaver was devastated, and a new taste for silk hats instead of beaver in the world cut into the demand for fur. White explorers and visitors apparently made little use of the area until the rush for gold began all over the West in the middle years of the 19th century. There were profitable mineral resources at the head of Soda Butte Creek and the Clark's Fork of the Yellowstone near present day Cooke City, and also along Bear Creek near the present town site of Jardine, Montana. Because most of Yellowstone Park's geology is from relatively recent volcanic activity, there are fortunately no significant deposits of valuable Gold Rush minerals within its boundaries. As for the physical beauty of the Park, its abundant wildlife, and its astounding geothermal features, human beings have known about these features of the Yellowstone Valley for at least 9,000 years. However, because the central Rocky Mountains offers many beautiful peaks and valleys, because wildlife suitable for hunting favors lower elevations, and also because geysers and hot springs counted as little more than oddities for Paleo-Indians and early Euro-American explorers, the Yellowstone region was not targeted as a place for easy living and profit.

*Devil's Thumb Soda Deposit,* F. Jay Haynes black and white stereocard, No. 1009 (ca. 1886).

Visitors were particularly intrigued by the Devil's Thumb and Liberty Cap travertine domes at Mammoth. They have been inactive since the first written accounts. P.W. Norris, the second superintendent of Yellowstone, apparently wanted to reactivate these features by piping water into the tops. This stereocard shows the end of one of these delivery pipes. It is impossible to grow a dome by delivering water from the top. Instead, domes are built from the bottom up in the rare event when a spring bubbles to the surface within a narrow range of velocities that allows the water to spread out uniformly in all directions. Architects still want to mimic these beautiful features at Mammoth and grow living geology on buildings. This process is so difficult to reproduce and expensive to maintain that it has rarely been achieved.

LUDWIG HOHLWEIN
MÜNCHEN
YELLOWSTON

Six-horse tallyho stages were used to take visitors back and forth from Mammoth Hot Springs to the train depot at Cinnabar. In 1903, the Northern Pacific Railroad extended the line to Gardiner and built a new log railroad station next to the recently erected Roosevelt Arch. The tally-ho coaches held 32 passengers on this dusty 6-mile trip up and down the hill to Mammoth. They transferred at Mammoth to smaller 4-horse stages or buggies for the remainder of their journey around the interior of the Park. This lithograph poster, commissioned in 1910 by Harry Child for his Yellowstone Park Transportation Company, was made by the Austrian artist Ludwig Hohlwein. This is Hohlwein's only known American piece (24cm x 90cm).

Right: *Pulpit Terrace Coating Springs*, T.W. Ingersoll, [St. Paul] black and white stereocard, 1886. Stereocards like these were originally sold as souvenirs. Now they are used for historic and scientific reference material to document changes in the Park over time.

Below: Baggage sticker for Yellowstone Park Transportation Company (7.5cm x 11cm, *ca.* 1900).

## YELLOWSTONE NATIONAL PARK—BIRTH OF AN IDEA

Opposite: Photographer William Henry Jackson returned to Yellowstone in the late 1890s to photograph the Park. This chromolithograph #53321 of Jupiter Terrace and the National Hotel at Mammoth Hot Springs was made to be sold as a souvenir. Many of Jackson's black and white photographs were hand colored and reproduced as chromolithographs with a lost asphalt color lithographic process by a company called the Detroit Publishing Company. Jackson was the manager of the company for nearly 25 years. They also produced an extensive Yellowstone Park postcard collection from many of Jackson's photographs (23cm x 17cm, 1898).

In 1870, an exploration of the region, led by Henry Washburn and Nathaniel Langford, assisted by a U.S. Army detachment, initiated a movement to designate and develop the Yellowstone for a new and special use: large-scale tourism, an unimaginable prospect before the coming of railroads. As the rail system spread deeper into the West, journeys that had required months of dangerous and difficult travel could be accomplished with relative safety in just a few days. A new concept appeared in 1872, when legislation signed by President Ulysses S. Grant "set aside" part of the Yellowstone plateau as the world's very first National Park. The idea was to save this vast area of wilderness from the usual fate, ownership and exploitation by private interests, and to offer it protection for the enjoyment of the general public.

FOR THE
ENJOYMENT

A patch of sunshine in a dark, rainy sky creates a rainbow over the Roosevelt Arch at Gardiner, Montana.

Most of the stars we see with the naked eye are our neighbors in our galactic hometown, the Milky Way, a spiral disc of nearly 200 billion stars spinning around a black hole. Our home galaxy is a small one in a clump of about 25 galaxies called the Local Group. It is estimated that there are more stars in the universe than there are grains of sand on the Earth.

# WINDOW ON THE UNIVERSE

## SCIENCE, ART AND THE AMERICAN WEST

THE SPIRIT OF THE AMERICAN WEST, as a vitally important dimension of our national character and identity, encompasses both the best and the worst of our history and collective experience. Rugged courage, fearless independence, and enthusiasm for the spectacular landscape, all of these virtues were shadowed for decades by ruthless exploitation and senseless destruction. Nonetheless, a reverence for this vast region, and sound thinking about how to protect it, have endured and deepened since the end of the 19th century.

In his 1871 survey, Hayden wanted to find that rare phenomenon of geysers that Bridger had described to him twelve years before. At Fort Ellis, near Bozeman, Montana, Hayden's party joined with a mapping reconnaissance group led by John W. Barlow and David P. Heap, two officers in the U.S. Army Corps of Engineers. The two expeditions traveled with or near each other for several weeks, returning to Fort Ellis near the beginning of September in 1871. The information they gathered about the amazing Yellowstone Plateau was instrumental in convincing Congress to set aside the area as a park under the control of the U.S. Federal Government.

After these initial forays, scientific interest in the wildlife of Yellowstone continued and constantly grew. Among the early studies was Edward Warren's *Life of the Yellowstone Beaver* (1922 and 1926), unprecedented in its record of the skills and habits of a mammal that at one time had been trapped almost to extinction in the continental United States. Wildlife research in Yellowstone today includes projects to learn more about the large charismatic animals (wolves, grizzly bears, elk, bison), and smaller creatures (salamanders, bats, insects). One ongoing study that often shows up on television and in newspapers focuses on the reintroduction of wolves into the Park's ecosystem with a starter population of 31 released in 1995 and 1996. There is an irony here, of course, that for many generations, local people have understood dimensions of the natural world that have only recently been gathered into scientific studies and data-sets: the rocks and minerals underfoot, the ways of wildlife, and patterns of the weather. Though modern science provides us with detailed and profound explanations and insights, wisdom gathered by our forebears from direct experience should never be discounted in this long history of wanting to know how Yellowstone works.

*Walk away quietly, in all directions, and taste the freedom of the mountaineer. Climb the mountains and get their good tidings. Nature's peace will flow into you as sunshine flows into trees. The winds will blow their own freshness into you and the storms their energy, while cares will drop off like autumn leaves.*

—JOHN MUIR, 1838-1914

Opposite: Bull moose feeding on underwater vegetation in Beaverdam Creek.

Below: Ravens are one of the smartest birds in the world and prosper throughout most of the northern hemisphere. They have been observed flying in the "Death Zone" on Mt. Everest above 8,000 m and along the edge of the sea in Alaska.

Grizzly bear sow with her two six month old cubs looking for food in the sagebrush steppe habitat of Swan Lake Flat.

## LIFE IN EXTREME ENVIRONMENTS

The establishment of the National Park Service in 1916 was a watershed event in the support and advancement of scientific investigations in Yellowstone during the 20th century. Though he was well along in years by then, John Muir, the famous naturalist and founder of the Sierra Club, was playing a vital role in making this happen. Having published for decades about the wonders of the West, his celebration of the "Range of Light" in wilderness settings had caught the national imagination. He had been involved, after the turn of the 20th century, in three timely meetings that made a great difference in conservation. In 1903, President Theodore Roosevelt invited Muir to join him to "talk freely around a campfire" in Yosemite, to "drop politics absolutely" and discuss together their views and hopes regarding the American wilderness. Those conversations intensified Roosevelt's passion for conserving and understanding the natural world in which he had for so long delighted to hunt. Roosevelt's dedication thereafter to the creation and support of our national parks is commemorated by the tall columnar basalt stone arch named after him, which marks the north entrance to the Park just a few kilometers below Mammoth, near where the Gardner River flows into the Yellowstone.

These pronghorn antelope are feeding on sagebrush and grasses during a light snowstorm. There were an estimated 100 million pronghorn on the North American continent 500 years ago. Like the bison, pronghorn were hunted almost to extinction, but have made a good recovery in part due to the security of places like Yellowstone.

The second crucial meeting involving Muir was in 1912, when he accompanied Stephen Mather, a wealthy and influential industrialist and ardent member of the Sierra Club, on a hike through Sequoia National Park, which had been created by an Act of Congress about fifteen years before. Soon after his trek with Muir, Mather vowed to intervene and end the poor oversight and misguided governance that he had witnessed with regard to our national parks. Subsequently, Mather was invited by Secretary of the Interior Franklin Lane to come to Washington D.C. and create a cohesive federal oversight agency that is the National Park Service. Mather became the first Director of the National Park Service, a post he held effectively and enthusiastically for a dozen years until his death in 1930.

Muir's third important one-to-one connection was with Horace Albright when Albright was still a student at Berkeley. Energized by the encounter, Albright, after graduating from law school, became a legal assistant to Mather in the new National Park Service. Soon after, he became the superintendent at Yellowstone, and ultimately the second Director of the National Park Service. In that capacity, the friendship he nurtured with the philanthropist John D. Rockefeller Jr. led to wonderful benefits for the nation. Rockefeller purchased and donated 33,000 acres to the Grand Teton National Park, as well as much of the land that became Acadia National Park and Great Smoky Mountains National Park, and financed the rescue of crucial forests around Yosemite. He is celebrated by the John D. Rockefeller Memorial Parkway, the roadway connecting Grand Teton and Yellowstone Parks. We honor Horace Albright today with the Visitor Center named for him at Mammoth, the Yellowstone National Park headquarters.

hence they believed that the hottest springs in Yellowstone were sterile and devoid of life. However, in 1965, Brock and his colleague Professor Hudson Freeze from Indiana University discovered that the pink filamentous bacterium ***Thermus aquaticus*** (*Taq* for short) thrives at temperatures of 80°C (158°F). From samples collected in the outflow channels of Octopus Springs, they found a biomolecule, a ***thermostable enzyme*** called ***Taq polymerase***, which controls the rate at which long chains of nucleic acids are synthesized into DNA and RNA. Carl Woese and his colleagues at Illinois, upon reading about Brock's discoveries, ventured into the Park to take samples for their own use (also from Octopus Springs), which confirmed the identification of the Archaea as the third domain of life. Three decades after Brock and Freeze's initial discovery, Kary Mullis shared the 1993 Nobel Prize in Chemistry for refining the use of *Taq* polymerase to replicate DNA and RNA under automated controlled conditions in the laboratory ***(polymerase chain reaction*** or PCR). These achievements now drive trillion dollar international biotechnology industries.

Norman Pace, who worked closely with Woese at Illinois and is now a professor at the University of Colorado, as well as many of his contemporaries and junior colleagues, have led the way in the search for microbial life in the extreme environments of Yellowstone and around the world. Their integration of polymerase chain reaction with the latest sequencing technologies has permitted microbial communities to be quickly and easily identified without the traditional requirement of laboratory cultivation. The result has been to exponentially increase our knowledge of the phylogenetic diversity and metabolic capabilities of the "unseen microbial majority" of life on our planet. ***Metagenomic*** analyses, now a core component of these types of studies, use statistics to piece together the entire nucleotide base-pair sequence of DNA ***(genome)*** in natural samples of microbial communities. Most recently, this work is being combined with analyses of RNA ***(transcriptomics)***, proteins ***(proteomics)***, compounds produced by metabolic activity ***(metabolomics)*** and many other components that make up all living cells. As a result, a comprehensive picture can now be established of the identity, evolutionary relatedness, metabolic activity, and ecology of microbial cells in the context of their natural environmental conditions. Among the many invaluable insights this work has yielded is that microbial life in environments like Mammoth has thrived for a long time and evolved rapidly due to dramatic changes in environmental conditions ***(steep gradients)***: temperature, acidity, nutrient availability, reduction-oxidation potentials, and flow velocity.

Yellowstone is a good place to live, yet winter is the limiting season. Bighorn sheep, like this ram, survive the winter by foraging for food like rabbitbrush, a plant they would normally not utilize. They lose weight because of the stress of searching in cold deep snow.

One spin-off from this work on Yellowstone's microbial ecology and evolution has been the rise of a global industry called ***bioprospecting***, the goal of which is to discover new biochemical compounds in nature ***(natural products)*** and to develop them for commercial applications in industry, medicine, energy, and the environment. One of the thousands of results from bioprospecting, in addition to the discovery of ***Taq*** polymerase, is the discovery and modification of a naturally-occurring detergent produced by thermophile microorganisms ***(surfactants)***. These molecules form small sudsy masses of bubbles on the surface of ponds at Mammoth when the wind blows, and they are now a key ingredient in the manufacture of commercial laundry detergents sold around the world.

Little brown bats feed on flying insects living in and around the Mammoth Hot Springs thermal complex.

However, bioprospecting has also brought serious and complex ethical and legal questions concerning the intellectual and financial ownership of the resulting natural products. This trouble is acute when the original samples have been collected from our national parks, from the lands of indigenous peoples, or from *Global Commons* areas around the world, such as deep seafloor that is beyond international jurisdiction and considered by many people as belonging to all humankind. In a precedent-setting 1997 agreement between the National Park Service and one biotechnology corporation, it was decided that the Park should be compensated for drugs or chemicals discovered in its thermophile microbes. Therefore, on your next hike through Yellowstone, think of the global reach of the science being conducted here and also consider the vital stewardship, protection, oversight and partnership that the National Park Service contributes to these discoveries. Many nations and indigenous peoples around the world now require similar legal structures and compensation for bioprospecting in their natural environments.

## MAMMOTH AS A WINDOW ON THE UNIVERSE

Society everywhere on Earth is now faced with solving urgent needs, including meaningful advances in environmental sustainability, human medicine, and the generation of energy. There is also a planet-wide urgency to search for life elsewhere in the cosmos, perhaps as a clue to the meaning of life itself. As a result, there is an unprecedented need for educated, curious, and good-hearted people to come together from all fields of science and art to meet these challenges. Ordinary citizens everywhere, enticed by direct experience with natural settings or moved by photographs of the world around us, must join the quest to understand the Earth through the arts as well as through cutting-edge science. For the survival of the planet, and also for our own survival on its surface, we need to know much more about how physical, chemical and biological components of natural systems have evolved to work together, and to incorporate that knowledge into our spiritual and metaphysical life.

Remarkably, Mammoth Hot Springs has also been a wellspring of scientific discovery, not only for how the hot springs themselves function on Earth and on other planets, but also for how other phenomena in natural environments have come into existence and operate. In the case of Mammoth, this adventure of discovery has taken us in many different directions, shedding light on histories and processes that can seem far removed from this landscape. For example, work done here has provided insights into how life interacts with the chemistry and flow of water to control $CaCO_3$ deposition, a process called ***biomineralization***. But we have also learned here how to reconstruct and interpret ancient life forms and environmental conditions, beginning from fossilized $CaCO_3$ mineral deposits that tell us what this world was like when the rock was originally formed. This kind of knowledge has proved essential to the field of astrobiology and the search for life on other planets, which will require the analysis of rock samples brought back on space missions from other worlds.

This tabular iceberg in the Antarctic Ocean was melting, tipping and rotating back exposing blue cathedral-like melt patterns that had formed underwater.

## Coral Reefs on a Warming Planet

Though Mammoth Hot Springs and tropical coral reef ecosystems can seem like dramatically different environments, a careful look reveals striking similarities and important scientific parallels. The spring water at Mammoth, a salty fluid, is surprisingly close in composition to seawater because it is the result of deep underground chemical reactions between marine limestone and water. Moreover, as we have noted previously, the travertine rock deposits all around us at Mammoth are composed of $CaCO_3$ in the mineralogical form of aragonite. This is the same mineral that corals use to precipitate and grow their skeletons in warm, shallow tropical seawater. In fact, the composition of aragonite needles in Mammoth travertine and in coral skeletons is virtually impossible to distinguish when viewed under a microscope. Even at a distance of several meters, travertine configurations seen from the Mammoth Hot Springs boardwalks are eerily similar to those seen in coral colonies while diving on tropical coral reefs. Equally amazing is the fact that some key characteristics of microbe communities at Mammoth resemble those of microbes inhabiting coral tissues, their sticky mucus coverings, and in the surrounding seawater.

Corals with hard exterior skeletons *(stony corals)* are closely related to soft corals *(gorgonions)*, sea anemones, jellyfish and hydroids, in that they all have stinging ***nematocycts***. These cells contain a small hollow tethered harpoon that explodes on contact to deliver a paralyzing toxin (a ***sting***) into unwitting victims. Small single-celled algae, called ***zooxanthellae***, invaded the tissue of coral like these more than 250 million years ago. This mutually beneficial *(symbiotic)* relationship provides oxygen and sugars, produced by zooxanthellae photosynthesis, to the coral. In equal measure, the zooxanthellae thrive in a safe and stable tissue environment where a steady supply of $CO_2$ is provided by coral respiration. This symbiosis requires that these coral colonies of *Orbicella faveolata* (appropriately called the "Mountainous Star Coral") live at water depths shallow enough for sunlight to penetrate, called the ***photic zone***, which is a key requirement for zooxanthellae to actively photosynthesize. The water is so shallow and clear, in fact, that the underside of waves can be seen at the top of the turquoise water column overlying the corals. Fish are a vital part of the ecosystem, feeding on the coral's algae-rich tissues and eating the prolific mucous covering secreted by the coral as a protective physical, chemical and biological shield. In the process, fish inadvertently scrape off, ingest and excrete small chunks of coral skeleton. Through the course of geological time and in combination with sea urchins, chitins and other organisms, this type of bioerosion causes more destruction of the reef tract than storms and hurricanes combined.

The coral *Orbicella faveolata* appears as soft and smooth mounds growing on the seafloor surrounding the southern Caribbean island of Curaçao. Yet in reality, coral is mostly a rock-hard stony skeleton with only a thin millimeter-thick outer layer of living tissue. These skeletons, called coral heads, are so massive and resistant that they cause large waves to break around them, splinter boat hulls and become readily fossilized and preserved in the geological rock record.

Coral tissue is organized into individual ***polyps***, each of which has its own mouth and tentacles. However, each polyp is also connected into a collectively shared nervous and digestive system. Corals therefore function both as individual polyps and as communal networks of colonial polyps. As a result, stony corals have a strategic competitive edge for survival in the nutrient-depleted surface waters of the tropics. This includes systematically changing the annual three-dimensional distribution and density of zooxanthellae, coral light protection cells *(chromatophores)*, and coral mucous production cells *(mucocytes)* within the polyp tissues.

Left: Individual coral polyps, each two millimeters in diameter, comprise the thin outer tissue surface of *Orbicella faveolata*. Green zooxanthellae are embedded around the mouth of each polyp. Brown pigmented tissue have fewer zooxanthellae. Eight translucent tentacles encircle each polyp mouth, which are in different stages of opening. Other polyps have completely retracted to form a protective brown dome.

Right: This one millimeter diameter laser-scanning microscopy image is of the lower-left quadrant of a single *Orbicella faveolata* polyp. The polyp mouth, in the upper right corner, transitions toward the lower left into green colored tentacles. These appear as small rounded nubs because the tentacles were retracted at the moment they were collected. The lobe-shaped masses are coral tissue that has relaxed and folded like an accordion *[modified from Sivaguru et al., 2011]*.

This shows that the coral animal utilizes multiple ecological strategies throughout the seasonal progression of each year. In the winter, sea surface temperature drops to 25°C to 28°C, which is low enough for the zooxanthellae to actively and efficiently photosynthesize. The coral is in an autotrophic feeding mode at this time, receiving the majority of it nutrients from algal photosynthesis. Conversely, in the summer months when sea surface temperature reaches a warm 29°C to 31°C, the zooxanthellae become stressed, and reduce or stop photosynthesis. They can even be ejected by the coral host into the overlying seawater during *bleaching* events if sea surface temperature is high enough. In these warm times of the year, the coral adopts a heterotrophic lifestyle by consuming previously formed particulate and dissolved organic matter and not relying on the zooxanthellae for nutrients.

Corals also have amazing talents when it comes to feeding themselves. Each polyp can switch-hit among fishing for particles of organic matter from the seawater using their tentacles, absorbing dissolved organic matter through its tissues, ingesting sugars produced by the zooxanthellae, or even consuming the algal symbionts themselves. When times are extremely tough, corals can ingest their own mucus, but at best this is a very short-term solution. In addition, corals reproduce both sexually and asexually, an ability that is almost incomprehensible to human beings. During sexual reproduction events that are closely tied to lunar cycles, the coral expels *(broadcasts)* millions of egg and sperm into the water column where they fertilize each other and form larvae. While the majority of these larvae perish and are eaten, some travel a short distance in the water column, and land on hard surfaces *(spatfall)*, and begin growing new coral colonies, as we see happening on the southern Caribbean reefs of Curaçao. A small number of larvae survive for longer time periods in the water column and thus are transported much longer distances before spatfalling to form new colonies on other distant reefs. In contrast, coral use asexual reproduction to increase the size of an individual coral head. The polyp mouth contains special cells that bud, permitting a new polyp to grow directly from the previous polyp's mouth.

When bacteria derived from human waste disposal are released into seawater, they cause an initial infection called black band disease. Corals have evolved chemical defense mechanisms to protect themselves from natural marine microbial communities. However, the initial human sewage-born bacterial infection weakens the coral, releasing nitrogen. The black band lesion forms as a result of a secondary infection caused when normal marine bacteria suddenly can attack and infect the area of tissue weakened by the first wave of infection. The marine cyanobacteria that are abundant in coral black band disease infections are closely related to the cyanobacteria found in the Pond and Proximal-Slope environments at Mammoth.

As dynamic marine ecosystems, coral reefs play a vital role in the physical, chemical, and biological health of our oceans. Skeletons of corals and other marine organisms accumulating on the seafloor over long expanses of geological time form sedimentary marine limestone deposits, products of ***biomineralization***. Around the Bahamas of the Caribbean Sea and the Maldives of the Indian Ocean, limestone accumulations have created tropical reefs that are kilometers in diameter and thickness, the largest biomineralization deposits on Earth.

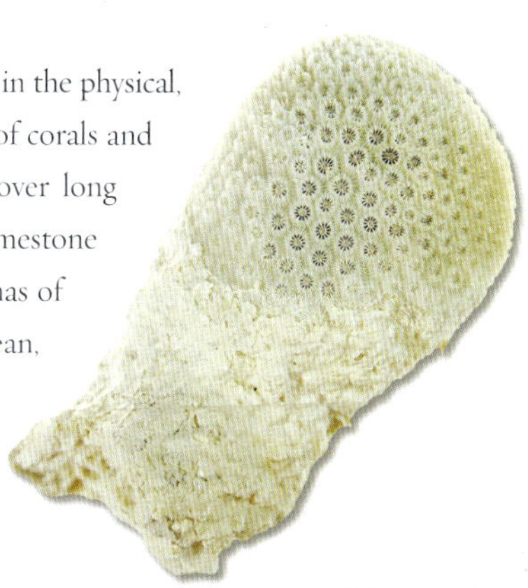

Left: A 10 cm long whole aragonite skeleton of the coral *Orbicella annularis* (a very closely related first cousin of *Orbicella faveolata*) after the tissue has been removed in dilute bleach. The upper right of the photograph is the aragonite crystalline skeleton that underlies healthy coral tissue and polyps, while the lower left is an older part of the colony that was encrusted with thin hard layers of calcifying algae. Each coral tissue polyp sits in a skeletal cup *(calyx)* with small thin vertical walls *(septa)* that radiate outward. The combined calyx and septa structure supporting each polyp is called a *corallite*.

What we have learned about the process of biomineralization at Mammoth helps us achieve new and better insights into how density banding (similar to tree rings) in the aragonite skeleton of tropical corals reflects the coral's response to changing sea surface temperature. X-ray diffraction (XRD) imaging of sliced coral skeletons reveals two fundamental processes that are active during the growth of coral heads. The first is represented by vertical to slightly inclined higher-density skeletal walls, which form around the margins of each calyx and extend upward as the coral asexually reproduces and buds new polyp clones atop each previous polyp.

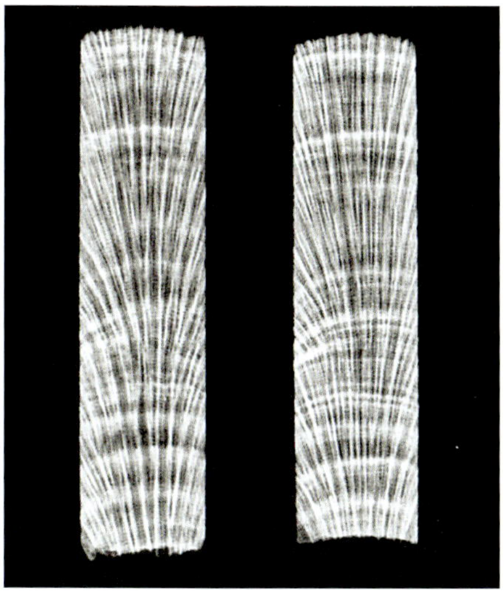

Bottom: X-ray diffraction *(XRD)* images of a 10 cm long *Orbicella annularis* coral head cut vertically down the middle into two flat rectangular slices, each one two millimeters thick. Dark vertical to inclined lines and horizontal layers are zones dense enough to prevent x-rays from penetrating, thus preventing exposure of the electronic x-ray film. Conversely, the white vertical to inclined lines and horizontal layers are areas where aragonite mineral density is low. Intermediate levels of skeleton density are shown by the many shades of gray in the x-radiograph.

The second is represented by concentric tree ring-like layers of alternating stacked skeletal bands that vary under XRD from black, to multiple shades of grey, to white. These are deposited by the coral during the complex interplay of annual changes in sea surface temperature, annual coral autotrophic-to-heterotrophic cycling, and biomineralization controls such as those observed at Mammoth that are derived from a combination of these abiotic and biotic influences. In general, the lighter less dense skeletal bands form when zooxanthellae photosynthesis is enhanced during the cooler winter seasons. This decreases the concentration of $CO_2$ and helps the coral to precipitate its aragonite skeleton in the same way that $CO_2$ degassing helps to drive travertine precipitation at Mammoth. The opposite is true for the warmer summer seasons, where reduced photosynthesis leads to slower growing and higher density skeletal layers.

One urgent issue associated with global warming is accurate prediction of how coral reef ecosystems will respond to three major environmental threats: rising sea surface temperature, a consequent rise in sea level driven by the melting of polar ice, and ocean acidification caused by increasing $CO_2$ concentrations in the atmosphere. Other environmental changes are impacting coral physiology and health. In many places around the world we are seeing a decline in vitality, an onset of bleaching, disease and other negative effects. The density bands laid down by coral over a span of years create a complex, well-preserved record of changes in sea surface temperature and the physiological response of the coral itself. The protein catalysis experiments completed at Mammoth allow us to sort out biological and non-biological processes recorded in coral skeleton banding records. However, the devil lies in the details for interpreting the intermediate gray shades of skeleton densities, which reflect the influence of changing sea surface temperature on zooxanthellae photosynthesis and coral biomineralization protein catalysis. Only in this case, the protein is created by coral cells rather than bacterial cells as it is at Mammoth. These insights permit changes in global sea surface temperature to be accurately reconstructed, and then fed into climate simulation computer programs, to aid in recommending sound policy strategies for combating climate change.

## Mammoth and the Mysteries of Roman Aqueduct Travertine

Beginning more than two thousand years ago and continuing for more than a thousand years, the civilization centered in ancient Rome constructed and maintained one of the largest, most ambitious water supply systems ever built by humankind. Eleven aqueducts serviced Rome itself, allowing the human population in and immediately surrounding the Eternal City to grow to more than a million inhabitants at the height of the Empire, rivaling the volumes of water supplied to support modern large urban centers. As water flowed along these aqueducts day after day and year after year, thick deposits of travertine, strikingly similar in composition and structure to the travertine deposited at Mammoth, were laid down along the floor and walls of the main channels in these great aqueduct structures.

The towering (10 meter high) aqueduct ruins at Romavecchia, an enduring testament to the accomplishments of ancient Roman engineering, have served for centuries as an awe-inspiring symbol of the powerful reach of the Empire. These supporting arches, strong and efficiently designed, show the dark, speckled volcanic tuff with which they were constructed. The Romans called this rock *peperino*, which is much like the volcanic tuff deposits in Yellowstone. The arcade in this photograph supports two aqueducts, the underlying Aqua Claudia and the Anio Novus, constructed from 38 to 52 AD to draw water from two different source regions. The Anio Novus, the largest and most ambitious of the Roman aqueducts, is rectangular in cross-section and constructed of Roman brickwork. The interior surface of the aqueduct channel was sealed with a layer of plaster that contained broken tile fragments, a mixture called *opus signinum*.

This travertine has been prized for centuries because of its distinct layering patterns for use in church altars throughout Rome. The entire research tool kit we use at Mammoth has been applied to the aqueduct travertine deposits, including the fieldwork, microscopy, and chemical and molecular analyses. Microscopic analyses of these types of solid rock require that a small rectangular sample, attached with epoxy to a glass slide, be cut to only 30 micrometers thick (one-third the thickness of a human hair), and is then polished to a mirror finish. Various types of light, lasers and energy can then be bombarded against, or moved into and through, the final glass-mounted rock slices *(thin sections)*. This allows the internal crystalline structure and chemical composition of the rock to be determined. This has permitted a sensitive reading of the ancient temperatures, flow rates, chemistry, microbial composition and even the hydrological manipulations by the Romans of the aqueduct waters at the time they were originally flowing.

The results achieved from aqueduct travertine analyses at three sites have been used to verify important historical and archaeological hypotheses about how the Romans regulated, maintained, and used these vital streams of aqueduct water. At Romavecchia, 8 kilometers southeast of downtown Rome,

Right: A 10 cm thick deposit of travertine collected from the base of the Anio Novus aqueduct channel at Romavecchia. Dark layers are rich in organic matter, including fossilized microbial mats (a version of stromatolite), and small crystals of $CaCO_3$ in the mineralogical form of calcite that is much like the Distal-Slope travertine at Mammoth. Lighter layers are composed of larger calcite crystals containing less organic matter. The layered centimeter-scale bottom-to-top stratigraphy of the travertine indicates that water velocity within the aqueduct increased over time. The finely laminated wavy layers in the bottom third are cross-sections of small current ripples that formed at lower water-flow velocities. The jagged void spaces and inclined layers are cross-sections of larger travertine ripples.

Bottom: A thin section of Anio Novus aqueduct travertine is 2 cm across and viewed with transmitted normal light. The prominent inclined and offset layering patterns are the cross-sections of small ripples. These patterns demonstrate that the millimeter-scale travertine layers seen in hand-held samples are themselves composed of an even finer Powers of 10 range of nanometer- to micrometer-scale organic matter-rich layers. These high-frequency, extremely thin layers, rich in dark organic matter, occur in many other types of rock deposits on Earth and might potentially be found in future rock sample returns from Mars and other planets.

well preserved remains exist of the Anio Novus aqueduct that collected water from the Aniene River in the Apennines and carried it 80 kilometers to the city, passing through the region of Hadrian's Villa and the famous Tivoli quarries. In samples collected from along its route, individual depositional layers within the travertine record the effects of rainstorms that occurred high in the Apennine range, as well as complex changes in the Italian climate. Other layers show signs of erosion, indicating times when the aqueduct was closed for maintenance or rebuilding. Each of these events changed the microbial communities being transported through the aqueduct, and their biomineralization effects, as recorded by the variety of extremely thin to thick layers seen preserved in the aqueduct channel travertine.

When in full service during the height of the Empire, these aqueducts were periodically scraped clean by armies of slaves using trowels and acetic acid (the molecule that gives vinegar its pungent taste and smell). During and after the Empire's fall, when this kind of maintenance came to an end, thick deposits of travertine continued to accumulate on the floors and walls of the water delivery system where the aqueducts were still in operation. Though earlier studies suggested that the Roman aqueducts were no longer in use much after 537 AD, when Ostrogoth armies from the east overran and devastated the city during the Gothic War, our isotope dating of the travertine deposits encrusting pipes and cisterns in the Baths of Caracalla reveal that aqueduct water flow continued for an additional 400 years, late into the first millennium. Furthermore, changes in the oxygen isotope composition of this travertine show that the aqueduct water was periodically warmed with large fires. This suggests that even in the Dark Ages, local people were still enjoying the Imperial pleasures of hot baths.

In the oil sands region of northern central Alberta, natural seeps of black oil emerge from small centimeter-sized vents in the olive-tan colored Ft. McMurray sandstone bedrock along the shores of the Athabasca River. Tenacious and adaptable orange microbial mats quickly metabolize the black streak of heavy oil. Understanding the metabolic mechanisms and environmental context for how these and other microbes break down hydrocarbons will eventually make us better stewards of the Earth's crust and its deep microbial biosphere, the largest reservoir of biodiversity on our planet.

Nonetheless, oil and gas *(hydrocarbon)* exploration and production will remain necessary into the foreseeable future, especially for the manufacture and use of plastics, as well as for heat generation. Therefore, because the global oil industry will be with us for a while longer, a vigorous search is underway for answers to large-scale environmental challenges. One of these is the need to enhance the production of oil that is thick and viscous, with a consistency akin to that of cold peanut butter *(unconventional hydrocarbons)*, because lightweight "sweet crude" oil, easier to extract, transport, and refine, is now becoming scarce. In non-destructive ways we need to make do with the hydrocarbon resources we have left. Much of this viscous oil is found in established fields where as much as 4/5ths of the available hydrocarbon resources may still be in the ground. One common mode of extraction is to pump seawater kilometers deep into buried rock formations and force the hydrocarbons out. Although pumped in under pressure, this does not reach the level of pressure that will break the rock and thus it is not *fracking*. Seawater is rich in dissolved sulfate ($SO_4^{2-}$), therefore it can nourish subsurface bacteria that produce deadly hydrogen sulfide ($H_2S$). This collateral damage effect, called *biosouring*, has caused a number of large offshore producing wells around the world to be closed *(shut in)* on an emergency basis in the midst of production.

Research conducted at Mammoth can help us address these issues. The same biomineralization process that creates layered mineral deposits from protein catalysis, which has proven so useful in other studies, is also applicable to these oil field challenges. This biological process, in which microbes can accelerate

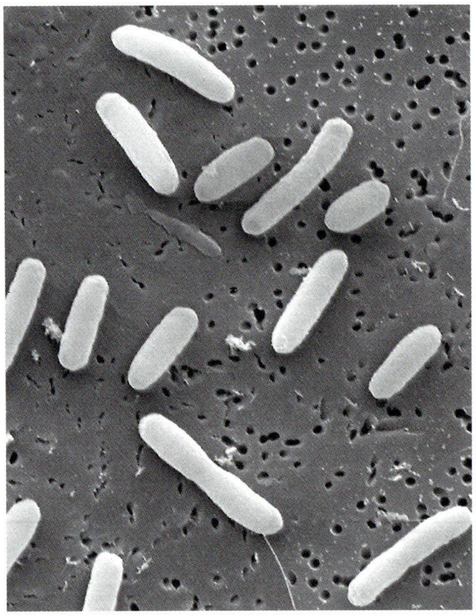

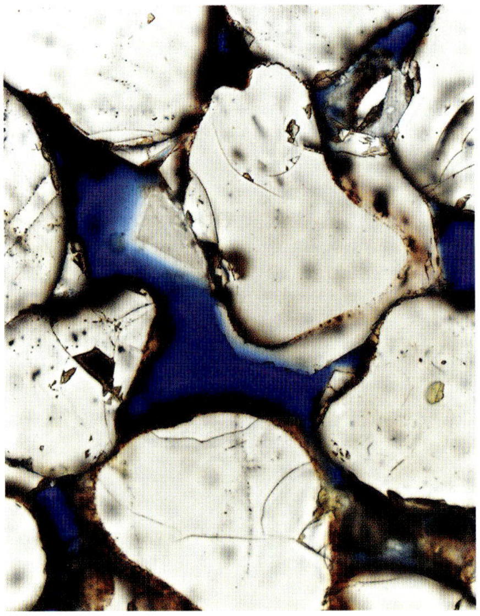

Top left: The thermophilic (heat-loving), halophilic (salt-loving) and piezophilic (pressure-loving) bacterium *Halomonas sulfidaeris* lives in pore waters trapped within the Cambrian-age Mt. Simon Sandstone at 2.2 kilometers underground. This scanning electron microscope (SEM) image shows the simple morphology of their one to two micrometer-long cells. The limited availability of nutrients requires that they have an amazingly slow metabolism, so slow in fact that they may clone their cells as infrequently as once every 100,000 years. Close relatives of *Halomonas sulfidaeris* thrive in deep hydrothermal vents of the Pacific Ocean and many other oceanic environments.

Top right: This three millimeter diameter field of view, thin section photograph of Cambrian-aged Mt. Simon Sandstone was collected from a well core, 2.2 kilometers underground in the Illinois Basin of North America. The blue is dyed epoxy injected to reveal the pore spaces between coarse rounded grains of quartz. Each sand grain is naturally coated with a dark red rind composed of iron oxide minerals *(hematite* and *goethite)* and microbial communities, all of which formed deep underground. These are environments that harbor the greatest volume and diversity of life on planet Earth.

$CaCO_3$ mineral growth, can be beneficially fostered in the pore spaces of reservoir *(hydrocarbon containing)* rock deep underground in an oil field. By reducing or even fully clogging porosity and permeability pathways of oil flow in rocks with $CaCO_3$ crystals, the efficiency of hydrocarbon flow along controlled flow paths can be enhanced. In addition, manipulations of temperature and fluid flow rates, based on what we observe in the natural life cycle of the springs at Mammoth, can reduce biosouring and the production of lethal $H_2S$.

## Mammoth and the NASA Astrobiology Roadmap

In cooperation with other national and private agencies around the world, NASA is opening a new era of space exploration and discovery. Multiple missions to planets, moons, asteroids, and even out beyond our solar system have boosted the acquisition of astounding new information about our planet, our solar system, and the universe out to its farthest reaches. This wave of exploration includes detailed studies of the Earth from space orbit, images of the farthest reaches of visible light with the Hubble Telescope, a close fly-by of Pluto, robot landers and rovers on Mars, and plans for rock sample returns and astronaut missions to Mars in the near future. Each mission has primary scientific goals, and astrobiology is one of these, studying the origin, evolution, distribution and future of life in the universe. The NASA Astrobiology Institute and the Astrobiology Roadmap have been established to answer three basic science questions: (1) how does life begin and evolve? (2) is there life elsewhere in the universe? and (3) what can we know about the possible future of life on Earth and beyond?

In research at Mammoth and other thermal springs in Yellowstone and elsewhere around the world, there is a special interest in understanding how life on the ancient Earth responded to changing planetary environments, and how these effects are manifest in long-term evolutionary change in microbial life. Furthermore, the ability to consistently identify and accurately read ancient biosignatures, the fossil remains of microbial life preserved in fossil travertine at Mammoth and other sites, will prove essential in efforts to find signs of life in rock samples eventually returned to Earth from other planets.

Instantly redefining the relevance of a "selfie," this 2015 self-portrait of NASA's Mars Curiosity uniquely captures the loneliness and harsh environs of the surface of the Red Planet. Among its many scientific missions, Curiosity completes extensive photographic coverage, as well as geological and chemical surveys that include the collection of drilled rock powder (see the light colored drilling debris at the lower center edge of the photograph). Rover and orbiter analyses have transformed every aspect of our understanding of Mars, ranging from acknowledgement that large bodies of water once covered major portions of the Mar's surface, to the detection of methane in the hunt for biosignatures and extraterrestrial life.

*Always continue walking a lot and loving nature, for that's the real way to learn to understand art better and better.*

—VINCENT VAN GOGH, 1853-1890

Astounding as they are, the famous images acquired by the NASA Hubble Telescope are not entirely unprecedented, nor our first dazzling contemplation of deep space, as artists took us there at the end of the 19th century. In 1889, Vincent Van Gogh, gazing from the barred window of his small room in the asylum at Saint Rémy de Provence in the south of France, completed the painting we know as ***The Starry Night***, now at the Museum of Modern Art in New York City. Although Van Gogh is often thought of as a visionary madman, we can tell from this painting that he was also a careful and dedicated observer. Indeed, in the tumultuous heavens he represents here, the largest swirling feature, directly in the middle of the canvas, is the M51 Whirlpool Galaxy, a star cluster first identified by William Parsons (later Lord Rosse) half a century before with an enormous reflecting telescope erected at Birr Castle in Ireland. We know for a fact that Van Gogh could see M51 from his window at the time he made this painting. Though he took many liberties with the galaxy's size and shape, he was one of the few artists to have ever shown us an actual galaxy as a prominent detailed feature in a painting. About other qualities of the universe he was truly prophetic: Van Gogh's skies over Saint Rémy are not embellished with peaceful little points of light, as one might see in fine art vistas from his own time or any previous era. He shows us instead a spectacular turmoil of creation and destruction, the immense and endless drama, uncannily like what we see in Hubble photographs now, more than a century after Van Gogh's death.

Opposite: *The Starry Night,, Vincent Van Gogh, Saint Rémy, June 1889, oil on canvas.*

Vincent Van Gogh's insights about the landscape around him, as expressed in *The Starry Night,* were radically new and different. His vision of the contrast between the bucolic familiar village below and the dynamic moving unknown storm of the universe seen in the night sky, was unique in 1889. Though few people admired his paintings or perceptions of the world when he was alive, his work is now widely revered, as we have come to better appreciate his vision—not as a wild dreamer, but as a careful observer of things as they really are.

Left: Now and then, some new technology emerges that can catch the interest and curiosity of a vast public. The NASA Hubble Space Telescope is such a breakthrough, framing the universe around us in a Powers of 10 context of mind-blowing proportions. At 31 million light years away and 60 thousand light years across, the M51 Whirlpool Galaxy, which Van Gogh featured in *The Starry Night,* is one of the brightest and most spectacular galaxies in the sky.

Water at 73°C emerges from a deep turquoise green pool at Mammoth. Wispy light beige filamentous bacteria thrive and form a shelf in the surrounding shallow water. Red, brown and green bacteria live in the cooler ponds beyond.

## EDUCATION IN AND ABOUT THE WILD

The search for truth in the natural world, and for the survival of our own species and so many others, requires us to open our eyes and minds, as scientists, artists, and ordinary citizens, to look carefully around us with the tools and strategies of scientific research, and also with the arts. Though we are obviously all in this together, so much of our daily life and educational experience is self-centered, a mentality that is both delusional and unsustainable. Central to learning about nature is to recognize that anything is possible, and that we cannot allow preconceived notions to warp and limit our encounters with new experiences. Without a measure of humility and a willingness to see the world with fresh eyes, accumulations of unchallenged "facts" can stop any further search for truth. More than ever before, our survival as a species requires us to give full and careful attention to our Earth, to where and what we are.

Photographs can provide a pathway to wonder. We want to *see*, and when we do, the mind begins to move and to search for the larger story. A photograph captures a bit of time as much as it captures location and light. It catches in a moment an ephemeral truth. Now, in a time of environmental crisis, is when such a sublime moment might be irretrievable, gone forever. In a photograph there can be realities about space and time that resonate with the continuum that Albert Einstein described more than a hundred years ago.

A partnership has been established between the NASA Astrobiology Institute, the University of Illinois at Urbana-Champaign, and local public schools in Montana and Illinois. Ongoing teaching and research activities include utilizing Mammoth Hot Springs as a universal laboratory to demonstrate the dynamic content and limitless possibilities of geobiology and astrobiology sciences. The spirit of curiosity, engagement, and fascination are naturally drawn out of these students by the wilderness at Mammoth, with the field experience requiring constant cooperation and care for one another. The only way to truly understand nature is to learn in its presence. Instantaneously, a young person's awe, admiration and closeness to nature are cast for life.

Undergraduate students from the University of Illinois at Urbana-Champaign participate in a yearly fieldtrip for research at Mammoth Hot Springs. Entitled *Yellowstone Biocomplexity and Astrobiology*, the seminar prepares students for several days of sustained fieldwork in the snowfall of late November, where they make predictions about and experimentally test microbe-water-travertine interactions, using time-lapse photography and infrared thermometers.

In some ways the natural world, as we encounter it now, is a snap-shot, a magnificent showcase for the unending dynamics of evolution, a narrative of mutation and change that can never again be quite the same as it is for us at one moment. So we encounter another paradox: the preservation of the natural world in its wild and living state is also the preservation of change, of metamorphosis. Only recently have conservationists appreciated this truth and reflected it in their advocacy and planning. Good nature photography is "mindful" in much the same way, showing us small and subtle changes in light and life, even as it shows us vistas that we feel are timeless. It recalls our experiences with Powers of 10, where even the smallest detail or story is mirrored or echoed in the vast truth in which we find it.

On this quest we try to make sense of the world, to understand as much as we can about where we come from and where we are going. The arts and scientific inquiry are great human legacies working together. The approaches and insights described in this book continue to help us understand the natural world better, to take care of it more wisely, to live in harmony with natural processes, and with each other. We strive to know, affirm, and celebrate our place in the universe.

Scientist collecting microbes, water and mineral samples at Mammoth Hot Springs under a formal research permit authorized by the National Park Service.

To nurture that all-important curiosity about the Earth, its remaining wild places, and its natural processes, requires that students go out to the springs, mountains, woods, grasslands and oceans. In these environments students can learn first-hand the complexity and immensity of these settings, the abiding mysteries of the wilderness, and the deep history and prospects of life on this planet. Working in such contexts can be hard; even the imperative to stand still, or sit quietly, and pay sustained attention can be difficult. Yet for this kind of intellectual and imaginative adventure, no comparable experience can be found back in lecture halls and labs. As you stroll past the hot springs and other famous sights in Yellowstone Park, you may see solitary people or small groups of them rooted to one spot. You may see them making photographs or notes with an electronic tablet or an old-fashioned paper notebook, or simply standing there with their hands in their pockets, gazing steadily at something, some small detail that others might only glimpse as they pass. Keeping still and looking carefully with an open mind can reawaken their intimate primordial connection with their home planet. Too many of us have forgotten that skill and that truth, something that all of us knew when we were children. For scientists, artists, and teachers, the challenge now is to help us all recover that connection.

> *Hold your hands out over the earth as over a flame. To all who love her, who open to her the doors of their veins, she gives of her strength, sustaining them with her own measureless tremor of dark life. Touch the earth, love the earth, honour the earth, her plains, her valleys, her hills, and her seas; rest your spirit in her solitary places. For the gifts of life are the earth's and they are given to all, and they are the songs of birds at daybreak, Orion and the Bear, and dawn seen over ocean from the beach.*
>
> —HENRY BESTON, 1888-1968

Steam from an Old Faithful eruption is more spectacular in winter because of the direct contact between the steam and –20°C air. Looking east toward the rising sun, the edges of the billowing steam were backlit and stood in colorful contrast to the landscape beyond. There is a silhouette of a bison cow absorbing the heat from the morning sun below the steam cloud. She would soon walk off to feed.

**EUCARYA PORTFOLIO**

At the time Darwin made his keen observations of the breathtaking diversity of the Eucarya, not much was even known about germs. He made the leap to see that changing Earth environments through geologic time influenced the emergence of new life forms through the process of natural selection. None of the Eucarya would exist without cyanobacteria photosynthesis and its creation of the envelope of oxygen that supports the entire modern day Earth's ecosystem. Woese's establishment of the Universal Tree of Life, which relied heavily on research in Yellowstone, provides a fundamentally new context with which to more fully understand the mechanisms by which the Eucarya have evolved.

Tom's photography in this portfolio pays homage to the remarkable variety of charismatic Eucarya around the world. This suite of photographs was chosen to provide examples of the three great lineages of the reptiles, all of which arose nearly simultaneously on the greenhouse Earth in the Late Carboniferous Period of the Paleozoic Era over 300 Mya. These groups include the Anapsids (turtles), the Diapsids (crocodiles, pterosaurs, plesiosaurs, ichthyosaurs, dinosaurs that eventually evolved into birds, and modern reptiles) and the Synapsids (reptiles that evolved into mammals). Initially the Synapsids rose to dominate the ecosystems of the Late Phanerozoic Eon.

Following the Permian-Triassic mass extinction event 252 Mya, when 85% of all life on Earth went extinct, the Diapsids proliferated and eventually ruled the terrestrial and marine environments of the Mesozoic Era for 186 million years. A meteor impact 66 Mya triggered the demise of the dinosaurs and allowed the Synapsid descendents to radiate and dominate the Cenozoic Era into the modern day.

Pictured at right:
The evolution of an amniotic egg was a way for reptiles to carry a watery and nutrient-rich universe onto the land. Turtles, crocodiles and birds still have this type of egg today. Turtles, likely the only surviving members of the Anapsids, sun themselves on rocks because they cannot internally regulate their body temperature, like most reptiles. This *Rhinoclemmys*, a neotropical wood turtle, near the west coast of Costa Rica, had a bee appear that seemed interested in the moisture on the tip of its nose. Annoyed by the bee, the turtle twisted its head back and forth before eventually retracting into the seclusion of its shell. The bee and the turtle are both members of the Eucarya. One has an external skeleton and the other has an internal skeleton and a shell. This interaction was between two closely related creatures within the same domain on the complex Universal Tree of Life.

Crocodiles and alligators, cold blooded egg laying reptiles, evolved early in the Mesozoic Era. They are ancient cousins of the dinosaurs and may be the least changed modern example of a Diapsid reptile. This Nile crocodile, *Crocodylus nilotichus*, is widely distributed in central, southern and eastern Africa. The name is derived from the Greek word for "pebble worm," referring to its beaded skin. This four meter long crocodile was photographed in September along the bank of the Mara River in Kenya, when wildebeests were migrating south into the Serengeti Plain. A crocodile's hunting strategy is to nearly submerge in the water to become almost invisible, then lunge and latch onto their prey, and drag them into the water. The pointed teeth allow them to hold their victims tight, drown them, and tear them apart.

Mammals evolved from Synapsid reptiles early in the Mesozoic Era by modifying the amniotic egg yolk into a placenta. *Gorilla beringei,* the mountain gorilla, is a critically endangered species found in central Africa, with less than 800 individuals known to be in existence. This adult female and her 5 week-old infant was photographed in Volcanoes National Park in northern Rwanda. They dwell on the slopes of five large volcanoes that make up the Virunga Mountains. Although close to the equator, these mountains are snow capped; therefore the mountain gorillas have much longer hair than their cousins, the lowland gorillas. They are extremely intelligent, social, and interact and play a great deal. The affectionate interactions between this mother and infant were the most endearing that Tom has ever witnessed in the wild. If humans were as caring to each other as these animals, we would have fewer problems in the world.